DISCLAIMER

Prepared Under DOE Contract # DE-AC05-00OR22725
Between
Department of Energy Oak Ridge Office
and
UT-Battelle, LLC

CLIMATE CHANGE AND
ENERGY SUPPLY AND USE

Technical Report for the U.S. Department of Energy in Support of the National Climate Assessment

Coordinating Lead Author: Tom Wilbanks, ORNL

Lead Authors: Dan Bilello, NREL
                              David Schmalzer, D. Schmalzer Services LLC
                              Mike Scott, PNNL

Contributing Authors: Doug Arent, NREL
                              Jim Buizer, University of Arizona
                              Helena Chum, NREL
                              Jan Dell, CH2MHill
                              Jae Edmonds, PNNL
                              Guido Franco, California Energy Commission
                              Russell Jones, API
                              Steve Rose, EPRI
                              Nikki Roy, Center for Climate and Energy
                              Solutions
                              Alan Sanstad, LBNL
                              Steve Seidel, Center for Climate and Energy
                              Solutions
                              John Weyant, Stanford University
                              Don Wuebbles, University of Illinois

Department of Energy   Bob Vallario
 Program Manager: Phone: (301) 903-5758
                              E-mail: bob.vallario@science.doe.gov

Process Coordinator: Sherry Wright, ORNL

CLIMATE CHANGE AND
ENERGY SUPPLY AND USE

TABLE OF CONTENTS

LIST OF FIGURES

LIST OF TABLES

CLIMATE CHANGE AND ENERGY SUPPLY AND USE

Technical Report for the U.S. Department of Energy in Support of the National Climate Assessment

Executive Summary

This Technical Report on "Climate Change and Energy Supply and Use" has been prepared for the U.S. Department of Energy by the Oak Ridge National Laboratory in support of the U.S. National Climate Assessment (NCA). Prepared on an accelerated schedule to fit time requirements for the NCA, it is a summary of the currently existing knowledge base on its topic, nested within a broader framing of issues and questions that need further attention in the longer run.

The report arrives at a number of "assessment findings," each associated with an evaluation of the level of consensus on that issue within the expert community, the volume of evidence available to support that judgment, and the section of the report that provides an explanation for the finding.

GCRP, 2009, indicates that the US energy sector is large and complex, with impressive financial and management resources, capable of responding to major challenges. It is accustomed to strategy development and operation in the fact of uncertainties and risks, both environmental and political. No sector has better capabilities to respond to challenges posed by climate change impacts.

Current knowledge indicates that such challenges tend to focus on climate-change-related episodic disruptions of energy supply and demand related to extreme weather events at a regional scale, on exposures related to risks in especially vulnerable areas, and on implications of changes in temperature and precipitation patterns – extremes as well as averages – for supply and use systems that are sensitive to climate parameters.

More specifically, the report's assessment findings are as follows. In each case, the report includes further information to support the finding.

Regarding implications for components of energy supply and use systems and cross-cutting implications for energy supply and use, we find that there are:

Implications for components of the nation's energy supply and use systems

- In most cases, the major current risk for both supply and use is from episodic disruptions related to extreme weather events

- Impacts from weather phenomena associated with climate change pose risks of economic costs to energy suppliers and users

- Increases in average temperatures and temperature extremes will mean increasing demand for electricity for cooling in every US region, along with reductions in energy demands for space heating

- Climate change is expected to have a larger impact on peak electricity demand than on monthly average electricity demand

- Impacts of climate change are risks to many oil and gas supply activities in vulnerable coastal areas, offshore production areas, and tundra areas

- Both climate change and rising concentrations of atmospheric carbon dioxide will affect bioenergy production potentials

- Expected seasonal and/or chronic water scarcity represent risks of electricity supply disruptions in many US regions

- Climate change will affect the geographical pattern of renewable energy supply potentials in the US

- Expected reductions in precipitation in the form of snowfall in the US West will reduce hydropower production, at least in some parts of the region

- In most cases, adaptation measures can reduce risks and prospects of negative consequences for energy supply and use

Cross-cutting implications for energy supply and demand

- Energy system resilience will benefit from progress with technology R&D

- Most vulnerabilities and risks for energy supply and demand reflect relatively fine-grained place-based differences in situations

- The variability of risks from weather-related events in both time and space will increase with climate change

- Climate change implications interact with and are affected by regulatory environments

- In many cases, gaps in the availability of data limit the capacity to answer key assessment questions

Regarding climate change risk management strategies for energy supply and use, we find that:

- Despite uncertainties about climate change impacts in the future, robust risk management strategies can be developed and – in an iterative manner that incorporates continuing observation, evaluation, and learning - implemented

- Many of the elements of such strategies can be identified based on existing knowledge

- A critically important step toward developing such strategies is conducting vulnerability assessments

Regarding knowledge and research gaps, we find that:

- Improving knowledge about vulnerabilities and possible risk management strategies is essential for effective climate change risk management in the energy sector

- Particularly important is improving knowledge about and improving capacities related to potentials for renewable energy development, resilience to extreme events, and potential tipping points for particular aspects of energy supply and use

Regarding the challenge of developing a self-sustained assessment process for the longer term, we find that:

- A self-sustaining long-term assessment process needs a commitment to improve the science base, working toward a vision of where things should be in the longer term

- Capacities for long-term assessments of vulnerabilities, risks, and impacts of climate change on energy supply and use will benefit from effective partnerships among a wide range of experts and stakeholders

- Self-sustaining assessment structures will provide value to all partners

REVIEW DRAFT

CLIMATE CHANGE AND ENERGY SUPPLY AND USE

Technical Report for the U.S. Department of Energy in Support of the National Climate Assessment

I.   Introduction

The third U.S. national assessment of climate change impacts and responses, the National Climate Assessment (NCA), will include a number of chapters summarizing impacts on sectors, sectoral cross-cuts, and regions. One of the sectoral chapters will be on the topic of *energy supply and use* implications of climate change in the U.S., as specified by the Global Change Research Act of 1990.

As a part of the NCA effort, a number of member agencies of the U.S. Global Change Research Program are providing technical input regarding the topics of the NCA chapters. For the *energy supply and use* topic, the U.S. Department of Energy (DOE) is the responsible agency; and this report has been prepared for DOE by the Oak Ridge National Laboratory (ORNL) in support of the NCA. DOE's interest grows out of a continuing research focus on climate change implications for energy supply and use systems, technologies, and services, as first demonstrated by its production of the US Climate Change Science Program's Synthesis and Assessment Product 4.5, *Effects of Climate Change on Energy Production and Use in the United States*, February 2008.

For broader issues related to relationships between energy infrastructures and others such as water and transportation, see a sectoral cross-cutting technical input report on Infrastructure, Urban Systems, and Vulnerability, also supported by DOE. For more attention to energy-water-land system interactions, see an additional technical report on that topic, supported by DOE,

All of the technical reports to the NCA are being prepared on a highly accelerated schedule. As an early step in organizing the NCA, a workshop was held in November 2010 to discuss sectoral and regional assessment activities. Out of that workshop came a number of further topical workshops and a working outline of the NCA 2012 report, including sectoral, regional, and cross-cutting chapters. In the summer of 2011, a number of USGCRP agencies stepped forward to commission technical input reports – each with at least one expert workshop and with a submission deadline of March 1, 2012, condensed into a period of eight months or less. Meanwhile, the advisory committee for the NCA (NCADAC) has appointed author groups for the report chapters, who will incorporate the technical input in a draft NCA report by mid-2012 for the first of several rounds of reviews and revisions, in order for the report to be submitted to the U.S. Congress in 2013 (see www.globalchange.gov). This report benefited significantly from an expert workshop co-hosted in Washington, DC, by the United States Energy Association (USEA) on November 29-30.

A final draft of the full report was sent to nine distinguished external reviewers, three of whom provided extensive comments and suggestions that were incorporated in this document. Other external reviewers provided supportive comments by telephone.

The report summarizes current knowledge, especially emerging findings since 2007, about implications of climate change for energy use, implications of climate change for energy production and supply (oil and gas, thermal electricity, renewable energy, integrated perspectives, and indirect impacts on energy systems), followed by discussions of implications for future risk management strategies, research gaps, and moving toward a self-sustained continuing assessment capacity for the longer term.

II.   **Background**

A.   *The Development Of The Report*

1)   Overview.

This technical input report is a summary of the currently existing knowledge base on climate change and energy supply and use, nested within a broader framing of the issues and questions that need further attention in the longer run. It builds on two previous assessments of implications of climate change for energy supply and use: SAP 4.5, February 2008, and pages 53-60 of USGCRP, *Global Climate Change Impacts in the United States*, June 2009, which were based largely on SAP 4.5. Its emphasis is on new knowledge that has emerged since SAP 4.5 went into document production in 2007.

2)   Approach.

The report was developed by an author team, led by ORNL, under the oversight of DOE, with significant input from a range of expert communities at the expert workshop on November 29-30. Data, methods, and tools depended on available source materials and varied according to the topic and the resources that have been invested in each particular topic. Judgments about report content, assessment findings, and levels of confidence reflect group consensus among the report authors, considering comments from selected external reviewers.

3) NCA guidance.

The NCA has adopted a range of types of guidance for the technical input reports covering eight topics that are priorities for the 2013 report: risk-based framing; confidence characterization and communication; documentation, information quality, and traceability; engagement, communications and evaluation; adaptation and mitigation; international context; scenarios; and sustained assessment (www.globalchange.gov/what-we-do/assessment/nca-activities/guidance). The ability to respond to this guidance was limited by several factors. First, the content of the report is based as much as possible on available sources of technical literature, which varied considerably in their treatment of such issues as scenarios and confidence characterization. In most cases, in fact, the sources do not refer to climate change scenarios at all. Second, the nature of much of the source material, often qualitative and issue-oriented, severely limited any attempt to estimate quantitative bounds on probabilities. And third, the highly compressed time schedule for the technical report preparation process limited potentials for engagement and communication and made it difficult to impose top-down strictures on report authors.

Given a body of source material that is a highly imperfect fit with the NCA guidance, this report has made an effort to frame its assessment findings in broad contexts of risk-based framing, scenarios, and confidence characterization. Assessment findings are associated with evaluations of the degree of scientific consensus and the strength of the available evidence. Where appropriate, findings are also associated with two general scenario-related framings of possible future climate changes: (1) "substantial, "which is approximated by IPCC Special Report on Emission Scenarios (SRES) emission scenario A2, and (2) "moderate," which is approximated by scenario B1.

4) Assessment findings.

Assessment findings are provided at the end of each major section of the paper, including the sections to follow on risk management strategies; knowledge, uncertainties, and research gaps; and developing a sustained capacity for continuing assessments. The complete list of twenty three assessment findings is included in this report's Executive Summary.

B.   *The Scope of the Report*

This report is intended as an update of the two previous energy assessments, considering energy sector vulnerabilities, impacts, and responses to climate change in the US, with additional attention to risk management strategies, research needs, and approaches toward a continuing national and regional assessment process. In line with other recent energy sector assessments, such as the UK Climate Change Risk Assessment (2011) and the World Bank report on Climate Impacts on Energy

Table 1. Energy sector vulnerability to climate change (World Bank, 2011)

	General	Relevant climate impacts		Impacts on the energy sector
		Specific	Additional	
Climate change impacts on resource endowment				
Hydropower	Runoff	Quantity (+/-); Seasonal flows, high & low flows	Erosion; Siltation	Reduced firm energy; Increased variability; Increased uncertainty
Wind power	Wind field characteristics, changes in wind resource	Changes in density, wind speed; Increased wind variability	Changes in vegetation (might change roughness and available wind)	Increased uncertainty
Biofuels	Crop response to climate change	Crop yield; Agro-ecological zones shift	Pests; Water demand; Drought, frost, fires, storms	Increased uncertainty; Increased frequency of extreme events
Solar power	Atmospheric transmissivity	Water content; Cloudiness; Cloud characteristics	Pollution/dust and humidity absorb part of the solar spectrum	Positive or negative impacts
Wave and tidal energy	Ocean climate	Wind field characteristics; No effect on tides	Strong nonlinearity between wind speed and wave power	Increased uncertainty; Increased frequency of extreme events
Climate change impacts on energy supply				
Hydropower	Water availability and seasonality	Water resource variability; Increased uncertainty of expected output	Impact on the grid; Wasting excessive generation; Extreme events	Increased uncertainty; Revision of system reliability; Revision of transmission needs
Wind power	Alteration in wind speed frequency distribution	Increased uncertainty of expected energy output	Short life span reduces risk associated with climate change; Extreme events	Increased uncertainty on energy output
Biofuels	Reduced transformation efficiency	High temperatures reduce thermal generation efficiency	Extreme events	Reduced energy generated; Increased uncertainty
Solar power	Reduced solar cell efficiency	Solar cell efficiency reduced by higher temperatures	Extreme events	Reduced energy generated; Increased uncertaint
Thermal power plants	Generation cycle efficiency; Cooling water availability	Reduced efficiency; Increased water needs, for example, during heat waves	Extreme events	Reduced energy generated; Increased uncertainty
Impacts on transmission, distribution, and transfers				

Table 1. Energy sector vulnerability to climate change (World Bank, 2011)

	Relevant climate impacts			Impacts on the energy sector
	General	Specific	Additional	
Transmission, distribution, and transfers	Increased frequency of extreme events Sea level rise	Wind and ice Landslides and flooding Coastal erosion, Sea level rise	Erosion and siltation Weather conditions that prevent transport	Increased vulnerability of existing assets
Impacts on design and operations				
Siting infrastructure	Sea level rise Increased extreme events	Flooding from sea level rising, coastal erosion Increased frequency of extreme events	Water availability Permafrost melting Geomorphodynamic equilibrium	Increased vulnerability of existing assets Increased demand for new good siting locations
Downtime and system bottlenecks	Extreme weather events	Impacts on isolated infrastructure Compound impacts on multiple assets in the energy system	Energy system not fully operational when community requires it the most	Increased vulnerability Reduced reliability Increased social pressure for better performance
Energy trade	Increased vulnerability to extreme events	Cold spells and heat waves	Increased stress on transmission, distribution, and transfer infrastructure	Increased uncertainty Increased peak demand on energy system
Impacts on energy demand				
Energy use	Increased demand for indoor cooling	Reduced growth in demand for heating Increased energy use for indoor cooling	Associated efficiency reduction with increased temperature	Increased demand and peak demand, taxing transmission and distribution systems
Other impacts				
Cross-sector impacts	Competition for water resources Competition for adequate siting locations	Conflicts in water allocation during stressed weather conditions Competition for good siting locations	Potential competition between energy and nonenergy crops for land and water resources	Increased vulnerability and uncertainty Increased costs

Systems (2011), it considers the entire range of climate change vulnerabilities, impacts, and adaptation potentials for energy supply and use (Table 1).

C.   *Emerging Trends And Contexts For Climate Change Implications For Energy Systems*

This report does not provide an overview of climate change expectations, possible socioeconomic patterns and trends affecting energy supply and use, global and national policy contexts, or broader issues for the energy sector itself, including current directions of technological change, although it considers these contexts in developing assessment findings. Representative references to climate change include NRC, 2010, and 2011 and are summarized elsewhere in the NCA report, as well as being incorporated in NCA guidance (see above). Socioeconomic trends and scenarios are also explored elsewhere in the NCA.

Global and national policy contexts are informed by such international efforts as IPCC and such national efforts as the NRC America's Climate Choices study (2010 and 2011). Energy options and choices, including issues related to technological change, are framed by such key references as NRC, *America's Energy Future*, 2009, and the ongoing work of the Integrated Assessment Modeling Consortium.

III.   Climate Change Implications For US Energy Supply And Use

This section summarizes current knowledge from research and practice about major vulnerabilities, risks, and impact concerns for different aspects of US energy supply and use in order to arrive at a number of summary assessment findings.

A.   *Implications Of Climate Change for Energy Use*

As the climate of the world changes, the consumption of energy in climate-sensitive sectors in the United States is expected to change. The most obvious and most-studied effects are changes in energy in buildings for space conditioning as a result of reduced demand for space cooling and increased demand for space cooling. Studies to date show regionally-varying decreases in the amount of energy expected to be consumed on site in residential, commercial, and industrial buildings for space heating, and increases for space cooling. Most studies project the effects of climate change and other variables affecting energy demand, but do not fully integrate all of the other factors affecting demand and supply or energy prices, all of which will affect actual future energy use. The following discussion emphasizes changes in demand resulting from climate change.

The current balance between energy use for heating and cooling in U.S. buildings varies by latitude (and to some degree by longitude) and can be expected to shift with warming from predominantly heating to predominantly cooling in some regions with moderate climates. Because the balance between heating and cooling differs by location, changes are expected in the balance of energy use among delivery forms and fuel types, as between electricity used for air conditioning and natural gas and fuel oil used for heating. Primary energy demand includes energy losses in generation, transmission, and distribution in both heating and cooling, but these losses are greater for cooling; so climate-change-induced switching from heating to cooling in regions with moderate climates tends to increase primary energy demand, even if site energy use declines. Increased cooling demand leads to increases in peak electricity demand in most regions, which increases the need to build electricity generation, transmission, and distribution facilities to meet the new peak (Miller et al., 2007, 2008;, Franco and Sandstad, 2008; Messner et al., 2009; Hamlet et al., 2010; Hayhoe et al., 2010; NPCC, 2010; Lu et al., 2010). It is likely that there will also be increases in energy (primarily electricity) used to pump water for irrigated agriculture and to pump and treat water for municipal uses. There is almost no new information concerning the impacts of climate change on energy consumption in other climate-sensitive sectors of the economy, such as transportation, construction, and agriculture. Although there are likely to be climate-change related decreases in energy used directly in certain processes such as residential, commercial, and industrial water heating, as well as increases in energy used for residential and commercial refrigeration and industrial process cooling (e.g., in thermal power plants or steel mills), there are no new studies documenting the extent of these potential changes. Since the publication of SAP 4.5, more new research has been going on internationally than in the United States (for a survey, see Mideska and Kallbekken, 2010). This section will focus on the United States.

1) Projections of energy consumption.

It is common for building energy demand projections to include temperature (often in the form of heating degree-days [HDD] and cooling degree-days [CDD]) as control variables to improve the precision of measurement in the income, price and other driver variables (Box 1). Analysts often investigate the implications of anomalously higher or lower temperatures as a sensitivity test of the projected robustness (e.g., ERCOT 2011. Analysts rarely investigate the impacts of systematic climate change on demand forecasts.

EIA (2005) investigated climate change impacts as side cases for the Annual Energy Outlook 2005. Warmer winters reduced residential and commercial building sectors' demands for space heating, which in turn reduced projected cumulative total fossil fuel use by 2.4%, but increased demand for space cooling and cumulative total electricity use over the forecast period by 0.2%. Sixty-two percent of fossil fuels consumption in buildings, but only 16% of electricity, was in temperature-sensitive loads. EIA followed with a side case for the 2008 Annual Energy Outlook (EIA, 2008) in which total building energy use fell by 2.4% and energy for electricity use increased by 0.7%. The net impact on total annual energy use (including effects of changing prices) was a very modest 0.4% . However, peak electricity use increased by 4.4 % in the summer while winter peak use fell by 0.8%. Summer electricity prices also increased for both residential and commercial customers.

Box 1: HDD and CDD Methods and Impacts

Since 2007, there has been additional recognition of the limitations of using CDD and HDD based on a building balance point of 65^0F (where the building is neither heated nor cooled) to estimate the effects of climate change on energy, particularly peak electricity. There has long been recognition that balance points differ between types of buildings and between regions (they tend to be lower for cooling and higher for heating in the northern states, while the reverse is true in the southern states. Adjusting these balance points leads to lower estimates of heating savings in the north and higher estimates in the south, while cooling costs are increased in the north and lowered in the south. Several researchers beginning with Belzer et al. (1996), recognized "dead zones" between base points for heating and cooling and fuel switching in heating and have worked out ways to estimate the appropriate adjustments. For example, these effects have been incorporated into models by Shorr et al. (2009), Miller et al. (2008), and especially Hekkenberg et al. (2009). In part due to computational burden, most climate change assessments of energy demand do not use hourly temperature forecasts from climate models .

Regional Studies. Most regions are summer-peaking regions for electrical demand. The Pacific Northwest has an atypical *winter-peaking* electrical system (due to high market penetration of electric heating). Even here, though, projected shifts in the seasonality of water availability for hydropower combined with projected increases in summer demand to cause summer peak problems. The Northwest Power Planning and Conservation Council's 6th Northwest Conservation and Power Plan discussed climatic change, and (NPCC 2010) addressed the impact of climate change on electricity demand. The Council staff determined that a 2°F increase in average winter temperature (3°F at peak) would result in a 600 MW decrease in average electricity demand and a decrease in winter peak demand,of 1,000 MW. In summer, the corresponding increase in July average temperature of about 3 degrees resulted in a 1,000 MW increase in average monthly load and a 3,000 MW increase in peak summer load. Together with increased hydroelectric yields in winter and reduced hydroelectric yields in summer, the net load/resource balance increased 1,200 average megawatts in winter and decreased by 3,220 average megawatts in summer. Resource adequacy improved in winter and declined in summer. Similarly, Hamlet et al. (2010), combining effects of direct temperature change with increased market penetration of air conditioning and continued population growth, concluded that

"...the combined effects of population growth and warming are projected to increase heating energy demand overall (22– 23% for the 2020s, 35–42% for the 2040s, and 56–74% for the 2080s), warming results in reduced per capita heating demand. Residential cooling energy demand (currently less than one percent of residential demand) increases rapidly (both overall and per capita) to 4.8–9.1% of the total demand by the 2080s due to increasing population, cooling degree days, and air conditioning penetration."

In California there have been a number of studies of the impacts of climate change on the electricity sector, several of which were just coming out as SAP 4.5 was being written, and the results of which were included in Box 2.2 of that document ("California's Perspective on Climate Change"). The California Energy Commission (CEC) and a number of individual researchers in California (e.g., Miller et al., 2007, 2008; Franco and Sanstad, 2008) have continued the analysis of climate change and its effects on state energy consumption. Messner et al. (2009) specifically investigated the effects on electricity demand in San Diego, while Xu et al. (2009) and Vine (2008) specifically considered adaptive responses. Incorporating climate change impacts on temperature, the most recent CEC forecast documents report that

"......the projected impacts of climate change in the mid and high demand scenarios on peak demand for the five major planning areas and for the state as a whole. By 2022, statewide peak impacts reach over 400 MW in the mid demand case and around 650 MW in the high demand case (California Energy Commission, 2011).

California and the Pacific Northwest share generating resources by long-distance transmission lines. The Pacific Northwest hydropower supplies may be less available in California in the future (Markoff and Cullen 2008, Perez et al., 2009). Lu et al. (2010) have demonstrated the adverse impact of simultaneous warming across the Western Grid.

Practice elsewhere varies. For example, the Tennessee Valley Authority has not included the impact of climate change on energy demand in their needs for power analyses, but is now on record for developing a Climate Adaptation Plan by June 2012 and updating it periodically (TVA, 2011). State assessments of the impacts of climate change do not necessarily deal with changes in energy demand. For example, Washington's does (Washington DoE, 2009), but Wisconsin's (Wisconsin DNR, 2011) does not. Although the 2009 New York State Energy Plan (New York State Energy Planning Board, 2009) mentions changes in demand for energy, no quantitative assessment was done. However, the scope of the 2013 plan appears to anticipate a quantitative assessment (New York State Energy Planning Board, 2011).

2) Impacts of climate change on building energy consumption.

SAP 4.5 found that on an annual basis, the amount of energy demanded for heating falls and the amount of energy demanded for cooling rises as a result of climate change. Since 2007 there has been much more extensive use of the self-consistent climate scenarios developed for the IPCC Special Report on Emissions Scenarios (Nakićenović, et al., 2000). Also there has been extensive downscaling of these scenarios for use in energy projections and more use of detailed regional scenarios (e.g., Washington DoE 2009; Miller et al., 2007, 2008; Lu et al., 2010; Xu et al., 2009; Hayhoe, et al., 2010). Some authors have made use of whole-building engineering models that are more sophisticated in handling of the impacts of lighting and internal gains on the heating/cooling balance of buildings (e.g. Crowley 2008). However, newer studies of both residential and commercial energy demand studies tend to confirm findings of SAP 4.5.

One of the more innovative studies estimated a climate response curve for electricity in California was based on unique individual billing data from residential customers of California's private utilities, assigned to individual zip codes and weather stations (Aroonruengsawat and Auffhammer, 2009). The authors note that changes in per household electricity consumption from climate change are driven by two factors—the shape of the weather-electricity consumption relationship and the change in projected climate. The steepest increases in demand were projected to occur in what are now the high-temperature areas of the state—the Central Valley and southeast California. Aggregate demand was projected to increase from 9% to 17% by the middle the 21st century. A 30% price increase could cut that growth by 11 to 14 percentage points, leaving electricity use largely unchanged at mid-century. Illustrating the importance of population growth in comparison with climate change, population growth projections to mid-century produces demand increases from 41% to 42%.

Market penetration of air conditioning

Predicting or accounting for increased future climate change-related market penetration of air conditioning has become more common in investigations of the impacts of climate change on energy use in buildings. In SAP 4.5, there was only one study found that had explicitly dealt with the potential impact of climate change on the market penetration of air conditioning, Sailor and Pavlova, 2003. Several subsequent studies have either adopted the Sailor and Pavlova approach (e.g., Shorr et al., 2009, Hamlet et al., 2009) or have modified it (McNeil and Letschert , 2007, Isaac and van Vuuren, 2009). In all of these studies, the increased penetration of air conditioning exacerbates the effects of hotter temperatures on space cooling energy consumption and on peak electricity demand.

Generally speaking, these studies have not reported the impact on peak demand, although it is clear that summer peak demand would be exacerbated. Both Isaac and van Vuuren (2009) and Shorr et al. (2009) discuss the countervailing impact of increases in air conditioning efficiency.

Extreme weather and peak demand

Some studies published in 2007 and later have explicitly considered the impact of heat wave conditions on peak electricity demand. These studies show that average daily demand increases non-linearly as CDD increases, while peak demand in creases roughly in proportion to maximum daily temperature. In Chicago, Hayhoe et al. (2010) looked at 99th and 99.9th percentile 3-hr periods, which increased dramatically, although they did not derive a quantitative impact on peak demand. In California, there has been detailed investigation of the impact of days in the summer whose daily maximum is hotter than would be expected 90% of the time under existing climate (Miller et al., 2007, 2008) and on peak demand days (Franco and Sanstad, 2008). At mid-century, California peak demand was projected to grows slightly faster than annual electricity consumption (Franco and Sanstad, 2008). Peak electricity kWh are typically much more expensive to supply than average kWh, and high demand may strain the capabilities of the transmission and distribution system, leading to power loss events.

Impacts of urban sprawl, heat islands, and community form on heating and cooling

Some studies have attempted to estimate the formation and effect of urban heat islands on energy demand in buildings. The U.S. studies include Rong, 2006, Contreras, 2009, Crawley, 2008, Rosenzweig et al., 2006, 2009). Crawley estimated the impact of "low" (1° C) and "high" (5° C) impacts of urban heat island effects on small office buildings in Washington D.C. (moderate-temperature humid climate). The heat island effects were similar in size to those of climate changes. Efficient buildings were less influenced than were standard buildings by increases in temperature. This implies less "benefit" on the heating side of the ledger, since better-insulated buildings require less heating to begin with. However, there is also less "cost" on the cooling side of the ledger, since the interior of the building requires less active cooling as summer temperatures rise.

Rong (2006) estimated urban sprawl as a index created from a principal components analysis. She used household characteristics data from the American Housing Survey, 2000 Census of Population public use sample and Residential Energy Consumption Survey (RECS) to estimate county-level energy demand for large U.S. counties as a function of housing selection, CDD, HDD, energy price and other variables. She then modeled residential energy use as a function of urban sprawl, indirectly through the mediators of house type and house size. National average impacts on electricity consumption were -5% for HDD (with most regions negative), and +17% for CDD, (with most regions positive). The overall model was then used to project the effects of climate change (including urban heat index) on residential energy consumption. Rong showed a slight UHI savings in primary energy nationwide currently because of the national dominance of heating over cooling under current climate (the opposite was true in cooling-dominated states). In the future the UHI effect on energy was estimated to cost energy at the national level, as savings in cold states (currently in the range of 3% to 10%) give way to losses (currently about 1% in warm states). This is because cold states will turn warm and warm states will turn warmer; so proportionately more time and energy will be spent on cooling.

3) Factors affecting heating and cooling besides climate: demography.

Population growth is still considered to be the largest overall driver of energy demand increases in buildings, especially residential. However, other demographic factors are also important. The age of the occupants could become important as the U.S. population ages (Ruth, 2006, Rong, 2006, Tonn and Eisenberg, 2007, and Crawley, 2008). Although many of people over 65 are poor and may depend on lifeline rates and fuel subsidies, Tonn and Eisenberg (2007) point out that the population aged 65-84 (growing much faster than the overall population) is expected to more than double between 2000 and 2050, and that currently older people consume more energy per capita than other age groups (e.g. 2.5 times the heating, over 3 times the cooling, and similar large differences for other end uses).

4) Water heating and cooling in buildings and industry.

Water heating is a major source of energy consumption in buildings, reportedly accounting for 2.58 quadrillion Btu or 12.9% of all site energy use in buildings in 2008, and for 3..81 quadrillion Btu (quads) or 9.5% of all primary energy use in buildings (DOE, 2011). There should be savings in water heating as temperatures warm. However, no studies were found of the potential energy savings associated water heating in buildings or for water demand in the residential and commercial sectors. Likewise, some of the energy use in industrial processes involves the heating and cooling of water. The 2006 Manufacturing Energy Consumption Survey (EIA, 2009) reports that of the total 15.7 quads of fuel consumption in manufacturing, about 3.5 quads was direct use for either process heating or process cooling and refrigeration. Much of this industrial energy consumption undoubtedly involves heating or cooling of water and some direct heating or cooling that also would be affected by climate change. However, the literature survey conducted for this chapter did not identify any new studies of changes in industrial energy consumption associated with climate change. Applying the estimate from SAP 4.5, a 1°C increase in temperature would produce a 6.2% decline in industrial energy use (0.2 quads) for direct process use.

In SAP 4.5, climate change was expected to increase demand for energy used for water withdrawals and distribution; however, there was very limited information reported in 2007. Almost no information was reported on energy use in transportation or construction, as affected by climate. Impacts on existing fuel use other than electricity were believed to be "small."

12

5) Electricity demand for water pumping and treatment.

Additional literature reviewed for this study contains data on the effects of climate change energy use to pump and convey water for irrigation. In addition, there are now estimates of water use in buildings related to efficiency, not climate change. National and regional information exists to calculate electricity use for treating water to potable standards and to move and treat sewage (EPRI, 2002).

The additional references contain estimates of energy consumption for water withdrawal, distribution, and treatment (Tables 2 and 3). Some regions have very large water demands for irrigation that could increase, based on the higher

Table 2. Impacts of Climate Change on Energy Use in Irrigation

Location	Energy Consumption	Source
Nation	Qualitative—increases with warmer temperatures	SAP 4.5
Nation	Nation, year 2000: groundwater, 700kWh/million gallons; surface water, 300 kWh/million gallons	EPRI, 2002
California	+173 GWh for 466,00 ac-ft lost reservoir water (increased precipitation, but changed timing of release.)	Burt et al., 2003
California	Currently (2001), 10.6 TWh, 18 million therms natural gas.	Klein, 2005
California	In addition to original canal and lifting costs, on-farm energy use is 30kWh/ ac.-ft; standard sprinklers 284 kWh/ac.-ft.; water transportation from San Joaquin Delta to Southern California is 2500 kWh to 5000 kWh/ac.-ft. Cited lifting costs for groundwater vary from 175 kWh to 740 kWh/ac.-ft.	Cooley et al., 2008
Nation	Dollar costs in United States $57/1000 m^3 for new irrigated land; $371/1000 m^3 for existing irrigated land. No estimates of marginal energy use reported.	Fischer et al., 2007

evapotranspiration due to warmer temperatures. For example, Burt et al. (2003) performed a study on current and future energy requirements for irrigation in California, including the projected loss of water from the state's reservoir system due to changed timing of snowmelt and surface water runoff. It was assumed that the lost capacity would be made up with groundwater, with associated additional energy consumption. The study did not take into account increased evapotranspiration from fields nor increased evaporation from reservoir surfaces. The calculated water shortfall was 466,600 acre-feet, and the corresponding increase in groundwater pumping energy was 173 GWh or about 0.37 MWh per acre-foot. The issue is similar in other Western U.S. snow-fed irrigation regions (Vano et al., 2010), but the calculations of irrigation energy impacts have rarely been done.

In 2001, 19% of the California's overall electricity consumption (48 TWh) and 32% of the state's total natural gas consumption (4.3 billion therms) was used to move and treat water and wastewater. Of that, agricultural use was 10.6 TWh but only 18 million therms (Klein, 2005). Costs may be exceptionally high in California, because so much water is moved very long distances within the state.

The EPRI 2002 study could provide the basis for estimating energy impacts of changes in irrigation demand due to climate change if the amount of water needed and the source of the replacement water were known. EPRI 2002 has a comprehensive picture of U.S energy use for water supply, water treatment, and wastewater treatment for the early 2000s period. Some of this data applies specifically to self-supplied water use for irrigation and livestock. For groundwater pumping, EPRI suggests a value of about 700kWh/million gallons (0.185 kWh/m^3). For surface water EPRI assumed an average value of 300 kWh/million gallons (0.079 kWh/m^3).

Fischer et al. (2007) have estimated the impact of global warming on U.S. irrigation, based on U.N. Food and Agricultural Organization-derived water deficits for various IPCC SRES scenarios. The cost of providing irrigation to an additional hectare of land was $290/ha, or $57/1000m^3, which includes "cost of supplying water from different sources, investment in irrigation equipment, facilities, land improvement, and computer technology; maintenance and repair, and labor". Additionally, they estimated pumping and energy cost and/or water price, operation and maintenance, and labor at $371/m^3. Unfortunately, they did not report the amount of energy assumed to be used.

DOE 2011 reported that water use in buildings in 2005 in the United States was estimated at 39.6 billion gallons per day, which was about 10% of all water consumption in the United States. Between 27 billion and 39 billion kWh were consumed nationally to pump, treat, distribute and clean the water used in the buildings sector, or about 0.7 to 1 percent of national net electrical generation in that year. Water use in the buildings sector also reportedly grew by 27% between

Table 3. Energy Use through Non-Agricultural Water Use (Public Systems)

Location	Findings	Source
Nation	No findings	SAP 4.5
Nation	Significant diversity in size and age of water supply and treatment systems. National average reference case in 2050 shows about 112 kWh/capita, with surface water treatment at 1,406 kWh /million gallons, groundwater supply at 1,824 kWh/million gallons.	EPRI, 2002
Nation	955 kWh/million gallons for trickling filter systems;1,322 kWh/million gallons for activated sludge; 1,541 kWh/million gallons for advanced systems without nitrification; 1,911 kWh/million gallons for advanced systems with nitrification.	EPRI, 2002
Nation	For bottled water, energy cost is water treatment is 10-1600 kWh$_e$/million liters, or about 0.0001 and 0.02 MJ(th) l^{-1}, Embodied energy in bottled water is about 5.6 to 10.2 MJ(th) l^{-1} (Average energy cost for Southern California municipal utilities is about 3000 kWh$_e$/million liters or 0.03 MJ(th) l^{-1})	Gleick and Cooley, 2009
National	7% of U.S. energy use is for providing water and waste disposal. Treatment cost: varies from 0.24 kWh/m^3 to 0.83 kWh/m^3 , depending on size of plant and type of process. Desalination requires 1.5 kWh/m^3 to 15 kWh/ m^3 depending on whether the water is brackish or sea water.	Novotny, 2010
California	Range of energy consumption, energy used in marginal water supply, treatment, and distribution: Recycled water 17 MJ/m^3 to desalination 42 MJ/m^3. Corresponding electricity consumption: recycled water 2.14 kWh/yr/m^3; desalination 5.2 kWh/yr/m^3	Stokes and Hovarth, 2009
Texas	2.1 to 2.7 TWh of electricity for water systems and 1.8 to 2.0 TWh for wastewater systems statewide, Texas uses 595,000 megaliters (ML) of water annually, yielding 3529 kWh/ML to 4538/ML for water systems . Varies locally.	Stillwell et al , 2011

1985 and 2005 (DOE, 2011), but the literature review for this study did not find estimates of the impact of climate change on non-agricultural water demand.

A small number of studies provide data on the costs of withdrawing, pumping, and treating water, although they do not directly examine the impact of climate change on these costs (Table 3). For example, Novotny (2010) states that about 7% of all U.S. energy use is for water and wastewater treatment. One percent or more is used to transport water and wastewater. Novotny also notes that domestic indoor water use ranges from 242 L/capita/day for a household without water conservation to 136 L/capita/day for a household practicing water conservation. Landscaping and other outdoor uses, leaks, and swimming pools increase the total to 650 L/capita/day (Novotny, 2010). GAO (2011) notes that "the energy demands of the urban water cycle vary by location; therefore, consideration of location-specific and other factors is key to assessing the energy needs of the urban water lifecycle."

They go on to note that factors include the source and quality of the water, distance and topography for conveyance, age and condition of the system (especially leakage rates), and level and type of treatment, all of which can vary significantly over even short distances (GAO, 2011, Stillwell et al., 2011, Stokes and Hovarth, 2009, Cooley et al., 2007). However, consumption of energy for treating water and wastewater are approximately linear in the amount of water treated (Stillwell et al., 2011); so if sources of water and methods of treatment are constant, the additional energy consumption required for this purpose under a changed climate would be proportional to the amount of additional water required. In theory, climate change that raised the average temperature of the atmosphere would also raise water temperatures for surface water (See Section IIIB, 2), and might also increase water consumption in landscaping. In the case of California, Stokes and Hovarth (2009) calculated energy consumption for a number of options to meet population growth. However energy costs would be similar on a per-volume basis to meet climate change–related shortfalls in supply or climate change-related increases in demand. Stokes and Hovarth's most costly scenario, providing all of California's current water needs with desalination, would require as much as 52% of the state's electricity. Comparable and even more detailed U.S. values for unit electricity consumption are available for public water supplies, wastewater treatment facilities , and self-supply by end users (EPRI, 2002). Cooley et al. (2007) note that increased water consumption also drives additional wastewater treatment, which results in additional energy consumption.

6) Energy demand in other industries.

Climate change likely will affect energy consumption in a few other climate sensitive sectors, such as transportation and agriculture (non-irrigation uses). For example SAP 4.5 discussed increases air conditioning in transportation (personal cars and refrigerated vans) and additional needs for cooling in livestock and poultry operations. The literature review for this study did not find any new U.S. studies that estimated effects of climate change on energy use in transportation or agriculture.

7) Impacts of adaptation and mitigation actions.

Buildings can reduce their air-conditioning loads by insulation, shading, and modifications such as reflective rooftops (SAP 4.5; Rosenzweig et al., 2006, 2009; Scott et al., 2008; Jo et al., 2010; Levinson and Akbari, 2010), but the degree of offset to climate change is less frequently computed. In one example, Shorr et al. (2009) modeled the impact of energy efficiency activities and calculated impacts on electric energy consumption in three groups of Northeast states. In most of that region, heating energy savings, efficiency upgrades, and market responses to increased cost (including fuel switching) could more than offset the impacts of additional market penetration of air conditioning and higher CDDs. But, significantly for states with warmer climates, that was not true of the southernmost tier of the northeastern states. These states saw increases both in energy use and cost. For a general overview of adaptation approaches and prospects in California, see Vine (2011). Jo et al. (2010) modeled 677 buildings in Phoenix using U.S. DOE's EnergyPlus™ model in Phoenix under today's climate, increased the average rooftop albedo (reflectivity) and estimated an annual electricity savings of a 4.3% in average annual electricity use. Under today's climate, Levinson and Akbari (2010) noted cooling energy savings on prototype high-reflectance commercial roofs in 236 U.S. cities per ranging from 3.30 kWh/m^2 in Alaska to 7.69 kWh/m^2 in Arizona (5.02 kWh/m^2 nationwide); the corresponding heating energy penalty in natural gas consumption ranged from 0.003 therm/m^2 in Hawaii to 0.14 therm/m^2 in Wyoming (0.065 therm/m^2 nationwide).

Under current climate, Rosenzweig et al. (2006, 2009) estimated that a combination of tree planting and green roof cooling strategies could reduce peak electricity use in some New York City neighborhoods by as much as 2 to 3 percent.
Reducing the demand for water also reduces the demand for energy to withdraw water from the environment, convey it, treat it, distribute it, and gather, convey, and treat wastewater. This can be an adaptive response to increases in water demand related to climate change. Several authors have discussed the impacts of water efficiency on regional or national water consumption, but generally have studied the impacts in the context of constrained supplies in today's climate, not climate change, and have not necessarily computed the resulting impacts on energy consumption. Water savings have been calculated for California by several authors, including Gleick et al. (2003), Klein (2005), Cooley et al. (2008), and for Las Vegas (Cooley et al., 2007).

8) Conclusions.

Broadly speaking, the main conclusions of the SAP 4.5 report concerning the effects of climate change on the future demand for energy in buildings remain valid. The annual demand for heating energy likely will decline and the annual demand for cooling energy likely will increase. In the northern states, where heating currently predominates, the impact on heating will be greater than the impact on cooling and the net impact on energy demand will be an energy savings. In the southern states and in some mid-latitude states, increases in cooling will more than compensate for declines in heating and the net use of energy in buildings will increase. These effects persist for both the older climate scenarios and the newer scenarios used by the IPCC.

What has changed are some of the details. Studies published since 2007 have attempted to estimate the effects of climate change while taking into account complicating factors such as the increased purchase and utilization of air conditioning as temperatures increase; electrification of heating systems as warming climates make heat pumps more practical; differential impact of increasing internal heat gains from lighting and plug loads on heating and cooling loads; effects on building loads of urban sprawl and urban heat islands; and "graying" of the population. More studies have addressed increases in system peak electrical loads due to increased cooling. Expanding the electrical generation, transmission, and distribution system to meet additional peak electrical load is a major potential capital cost of warming, regardless of what happens with total energy consumption. Most of the detailed complicating factors mentioned above tend to increase cooling demand and reduce heating demand, thus compounding the effects of climate change alone.

Climate change also is expected to increase the demand for water in agriculture and, along with growth in the population and economy, put more demand pressure on existing sources of water supply. In turn, this increased demand pressure for water in most places will mean that more energy must be used in pumping and conveying water for irrigation (and sometimes urban supply) and that more water will be lost in conveyance, storage and power plant cooling. Growing human populations increasingly compete for more distant water of poorer and poorer quality, which with more water demand likely will mean that more water and more waste water will have to be treated more aggressively to achieve drinking water standards. That, in turn, takes more energy, mostly electricity. More quantitative information has become available on the energy cost of water demand and supply as climate changes. In some states with elaborate long-range irrigation and urban water distribution systems or deep groundwater sources, the energy costs of supplying water are substantial and have the prospect of becoming larger still.

More studies are paying attention to adaptive responses in efforts to reduce impacts of climate change on energy and water bills and the environment. Examples include low-E windows to reduce solar gain and cooling loads, carefully designed building lighting, mass, shading, orientation and lot placement to reduce cooling and heating requirements, and urban design to manage sprawl and heat island effects. Many of these adaptive responses are promoted as "greener" or "more sustainable" solutions because they also improve the environment by reducing carbon emissions and water use. In this way they also mitigate some of the climate change for which they are intended to adapt. Conversely, some energy savings and carbon mitigation policies such as building codes and efficiency standards for building equipment and appliances may also offset some of the impacts of climate change and will have adaptive as well as mitigation value. Also, regardless of the motivation, saving water reduces the demand for energy to move it and treat it.

Quantitative estimates of the impacts of climate change on energy consumption in climate-sensitive sectors such as construction (up or down, depending on whether the construction season is lengthened or shortened), agriculture (up or down, depending on the direction and sizes of the water, chemical, and machinery burden), tourism (up or down, depending on whether the season is shortened or lengthened), and transportation (depending on the difficulty of maintaining movement and the effects on he number of viable transportation days) remain scarce. There is more interest in estimating the impacts of climate change on these sectors as economic entities than there is in estimating the impacts of changes in these sectors on energy demand.

9) Assessment findings.

Assessment findings about implications of climate change for energy use in the U.S. are incorporated in section III C, merged with those regarding implications for energy supply systems.

B.   *Implications Of Climate Change For Energy Production And Supply*

Energy production and supply includes a number of sub-sectors that differ in institutional responsibilities, knowledge bases, and possible climate change vulnerabilities. In a number of cases, significant new knowledge has emerged since 2007/2008.

1) Oil and gas production and supply.

The first assessment of implications of climate change for energy supply and use in the United States, SAP 4.5, included very little about oil and gas production and supply other than indirect effects of climate policy. By the second assessment, GCRP, 2009, however, attention to vulnerable regions (Alaska and the Gulf Coast) and early adaptations (to vulnerabilities of coastal facilities to flooding) began to redress the imbalance in attention to risks and vulnerabilities.

Since those two assessments, careful analyses of ways in which oil and gas production and supply are at risk from climate change impacts have begun to appear: e.g., Dell, 2010, and Burkett, 2011. In recent years, Dell has led efforts within the oil and gas industry itself to consider reasons for concern about climate change impacts and possible adaptation strategies, rooted in an argument that adaptation can be approached from a value-chain perspective, providing cost-effective approaches for identifying strategies and actions for which a business case can be made.

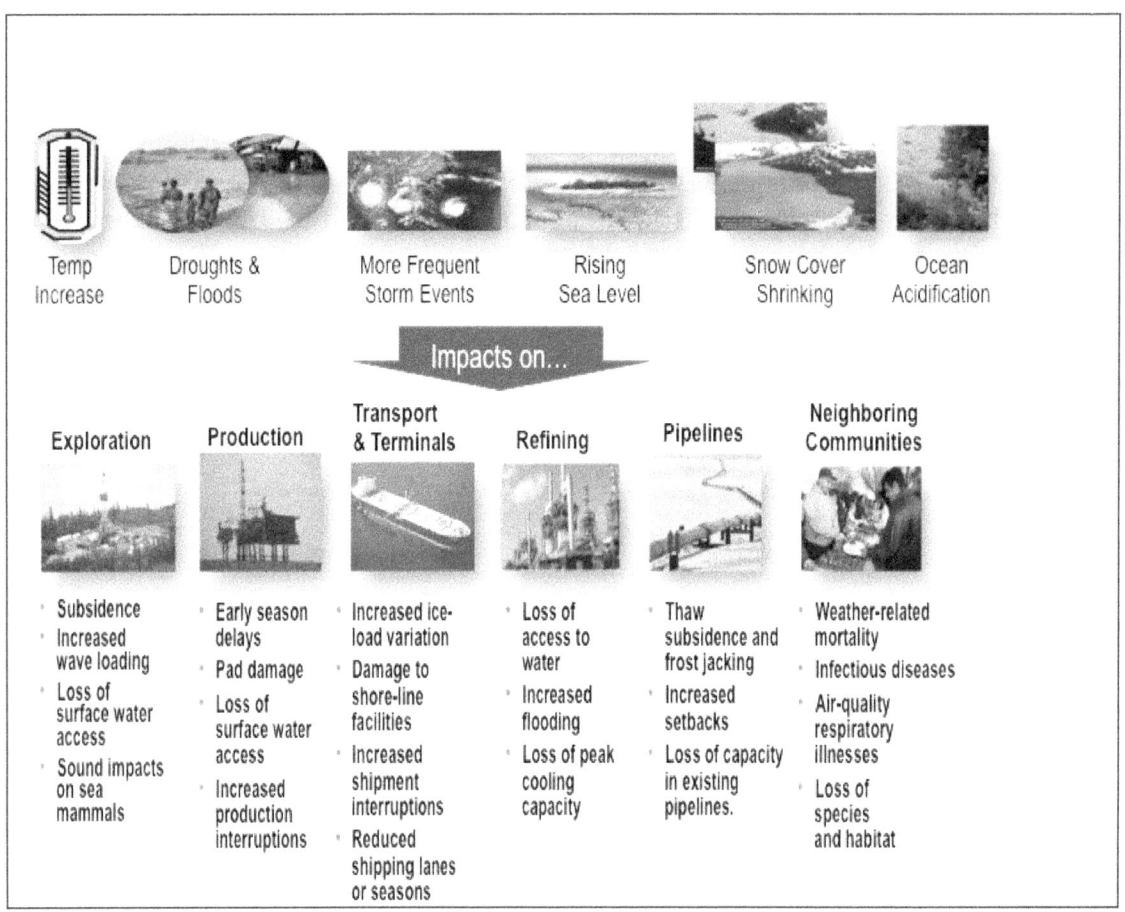

Figure 1. Summary of oil and gas sector impacts (Dell, 2010).

Clearly, the oil and gas industry conducts projects and operations in regions where temperature increases due to climate change will be especially severe (e.g., the Arctic), in areas affected by ocean acidification (off-shore production), in areas affected by sea-level rise and coastal storms (both off-shore and on-shore coastal areas), in areas affected by shrinking snow cover, and in activities that require significant amounts of fresh water to operate. This calls for scenario-based vulnerability and risk assessments as a basis for identifying opportunities and risks associated with adaptation strategy development. Dell's paper for the Society of Petroleum Engineers reports an evaluation of risks worldwide, considering two global impact categories and ten regional impact categories. Figure 1 summarizes the findings of this groundbreaking study.

Two case-study examples are especially instructive. In one case, the oil and gas industry has performed its own Alaskan Arctic Project Adaptation Assessment, looking at projected impacts of climate change on industry operations in Alaska (Dell, 2010). The assessment considered potential impacts on land-based infrastructure: the length of the season for tundra travel and winter construction (i.e., seasonal roads), permafrost as an active layer underlying buildings and transportation facilities, the length of the open water season, slope stability affecting pipeline corridors, snow depth affecting ice road construction, and onshore break-up patterns affecting delta river flooding. It also considered potential impacts on marine-based infrastructure: the length of the open water season (e.g., affecting exploration and construction seasons), the timing of freeze-up and break-up (affecting drilling seasons), the timing of fast ice formation and stability, storminess (affecting exploration drilling designs and downtime), and multi-year ice occurrence and thickness (related to design loads on structures). The assessment concluded that potentials for increases in five parameters due to climate change represented significant potential impacts: lightning strikes, tundra fire frequency and severity of polar bear encounters, coastal erosion and storm surges, and changes in permafrost which could impact piling design.

In a second case, Burkett (2011) analyzed climate change implications for coastal and offshore oil and gas development. The study identifies six key climate change drivers with the potential to both independently and cumulatively affect coastal and offshore oil and gas exploration, production, and transportation: changes in carbon dioxide levels and ocean acidity, air and water temperatures (especially in the Arctic), precipitation patterns and runoff (with potentials to cause difficulties in using coastal wetlands), the rate of sea-level rise, storm intensity, and wave regimes (threatening production platforms, bridge decks and supports, and pipelines). Figure 2 indicates interactions among physical climate change drivers affecting the coastal zone, many of which are already showing impacts of climate change. Other issues include effects of temperature increases and precipitation changes on oil and gas operations, especially water needs, where location matters a great deal in determining the degree of possible impact.

One way to view the importance of location for oil and gas sector vulnerabilities (suggested by Russell Jones) is to consider how the parts of the U.S. that are considered at greatest risk of temperature increases and precipitation changes (Figures 3 and 4) relate to patterns of oil and gas production. Overlaying these areas on the regions most important for U.S. domestic oil and gas production (Figures 5 and 6) suggests some possible issues associated with temperature changes in a high emissions scenario. Figure 7 indicates more significant vulnerability issues for ethanol production, where about 80% of current production comes from seven states that are subject to both precipitation and temperature changes. Issues may also exist for oil and gas production in the U.S from shale: e.g., water needs for shale gas fracturing.

Integrating climate-adapted bioenergy crops into agricultural and forestry landscapes, as agroecological zones shift, has the potential to avoid losses of production at a national level (Chum et al., 2011).

2)   Thermal electric power plant supply.

Thermal power plant supply is vulnerable to changes in water availability, greater frequency and duration of elevated regional ambient air and water temperatures, and increased frequency, intensity, and duration of extreme weather events (SAP 4.5). Considerable further work has been done since 2007 on these issues for thermal power plant supply, adding further understandings of risks and vulnerabilities.

Water availability

In some regions of the US, chronic or seasonal reductions in water supply due to decreases in precipitation and/or water from melting snowpack are likely to be significant, increasing the competition for water among various sectors including energy production (Kenny et al., 2009).

The production of energy from fossil fuels (coal, oil, and natural gas and also from nuclear power--is inextricably linked to the availability of adequate and sustainable

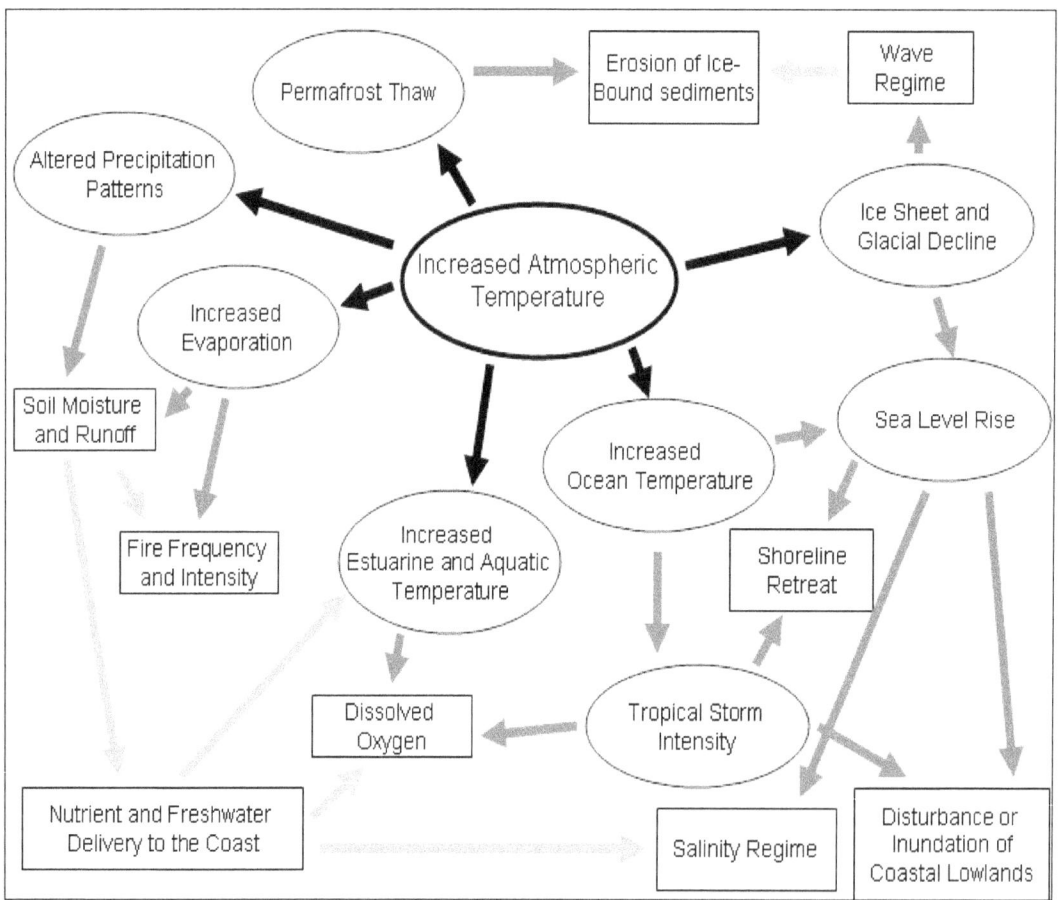

Figure 2. A conceptual model of the interactions among physical climate change drivers affecting the coastal zone. (Burkett, et al., 2009).

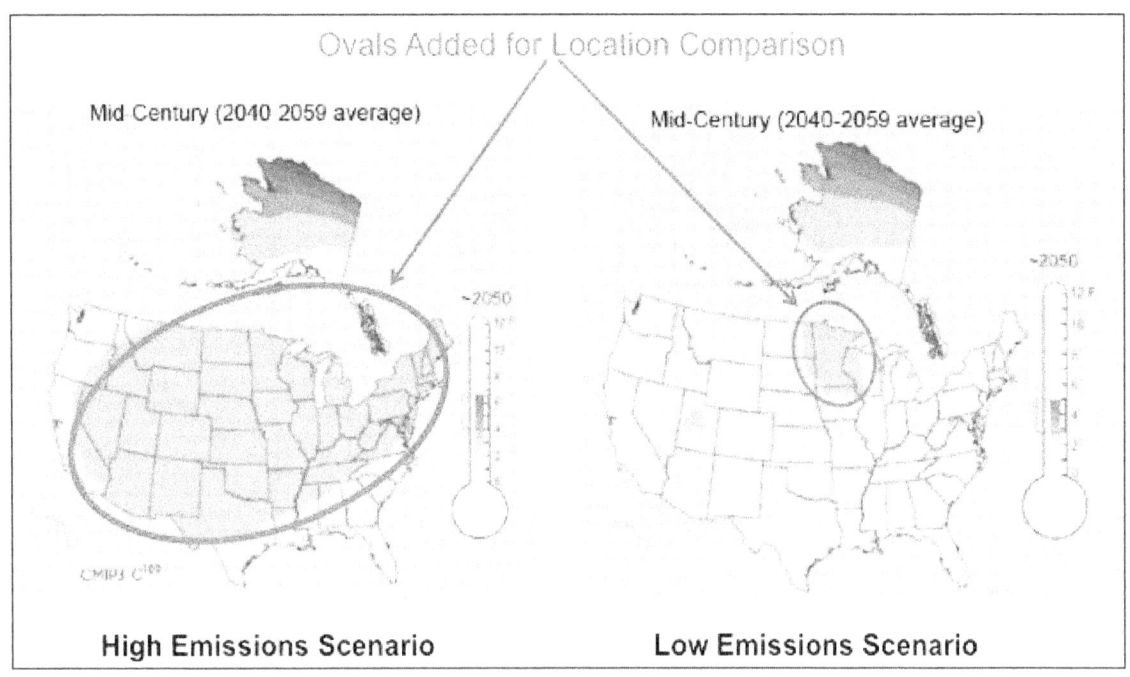

Figure 3. USGCRP: Potential Temperature Increases (from 1961-79)

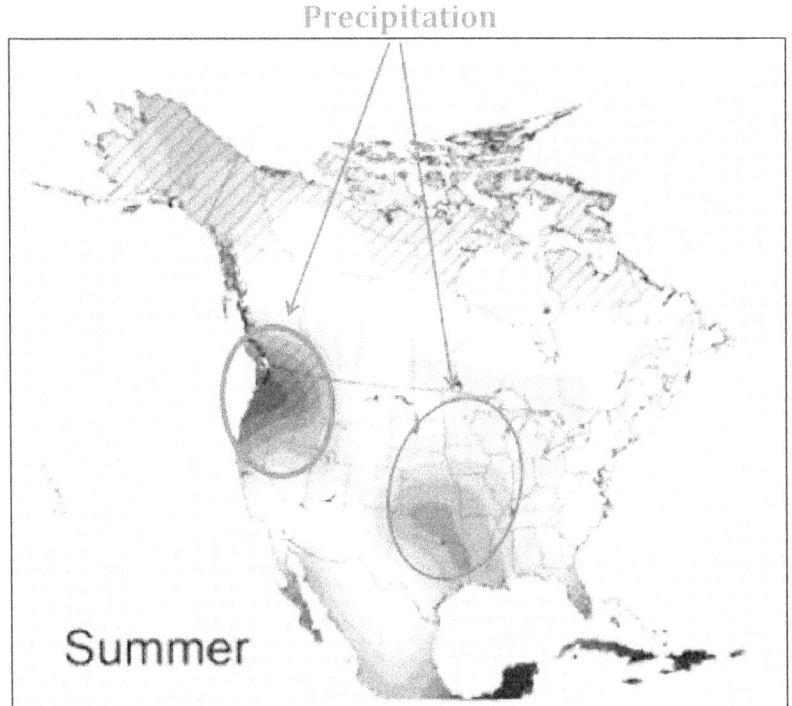

Figure 4. USGCRP: Potential Precipitation Changes by 2080-2099

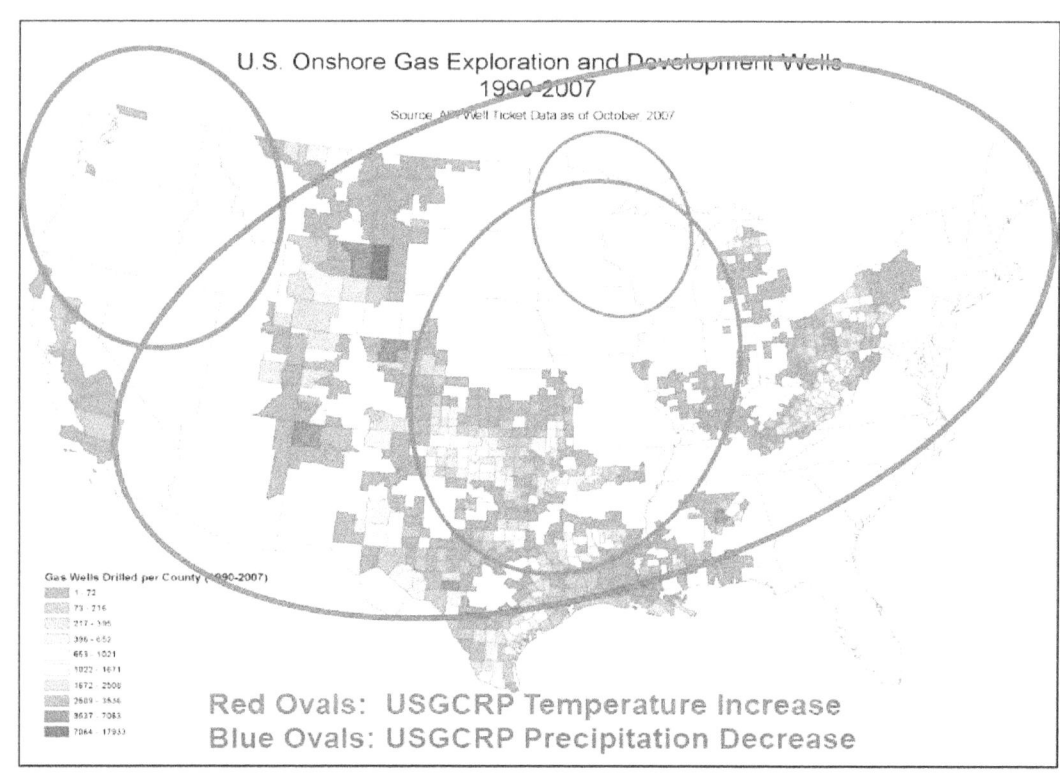

Figure 5. Lower-48 State Historic <u>Natural</u> <u>Gas</u> Development Areas

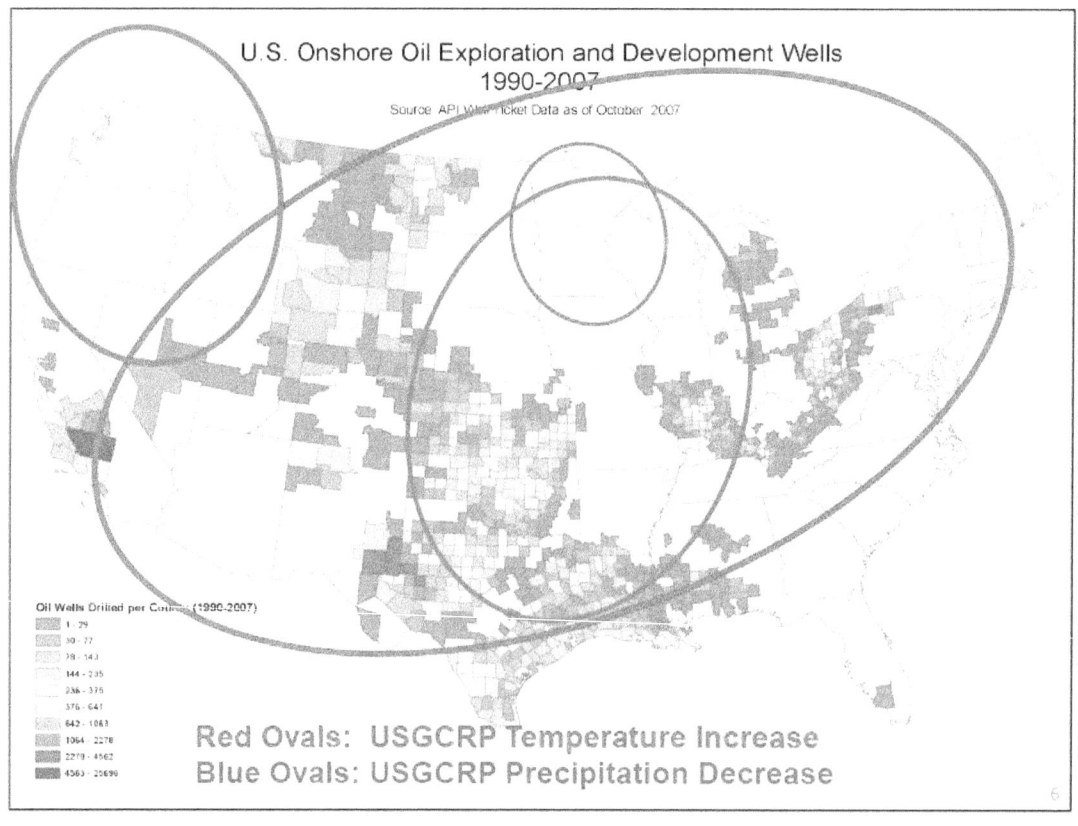

Figure 6. Lower 48 State Historic <u>Oil Development</u> Areas

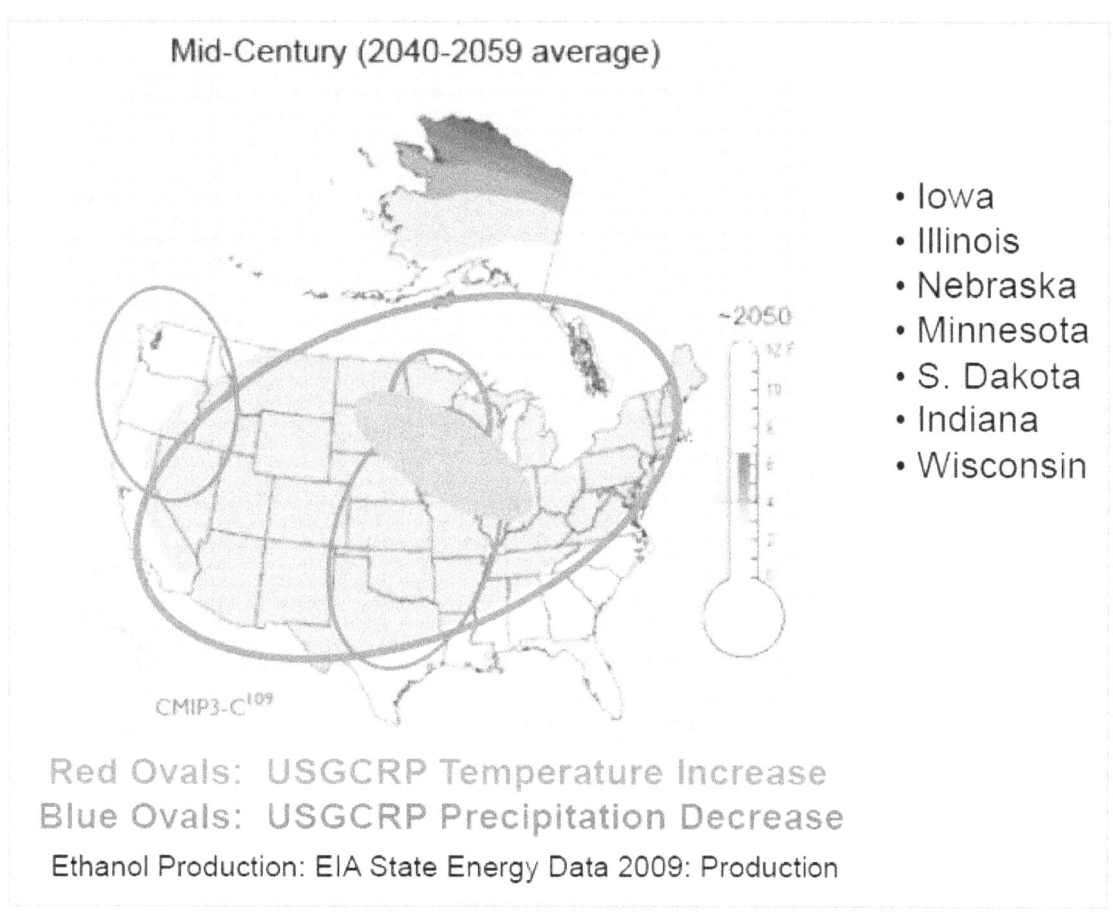

Figure 7. In 2009, 80% of US Ethanol Came from Seven States Potentially Subject to Temperature and Precipitation Changes (see green oval).

supplies of water (EPRI, 2011). While providing the United States with the majority of its annual energy needs, fossil fuels also significantly affects the nation's water resources in terms of both quantity and quality impacts (Cooley et al., 2011). In particular, the generation of electricity in thermal power plants (coal, nuclear, gas, or oil) is water dependent. In 2005, power plants were the largest source of freshwater withdrawals (41%), followed by irrigation (37%) (Kenny et al., 2009).

Studies conducted during 2011 indicate that there is a high likelihood that water shortages will limit power plant electricity production in many regions (See Box 2), pointing to growing regional water constraints, particularly in the Southwest, Southeast, as a result of chronic or seasonal drought, growing populations, and increasing demand for water for various uses, at least seasonally (UCS, 2011).

More specifically, the EPRI technical report includes scenario-based projections of water demand for 2030, related to drivers of demand rather than of supply. It finds that one-quarter of existing power generation facilities, or roughly 240,000 MW of

Box 2. Four Major Assessment Reports in Late 2011 Have Examined Water Use for Electricity Generation, Related to Concerns about Climate Change:

- Water Use for Electricity Generation and Other Sectors; Recent Changes (1985-2005) and Future Projections (2005-2030). EPRI Technical Report, November 2011
- Freshwater Use by U.S. Power Plants: Electricity's Thirst for a Precious Resource, Energy and Water in a Warming World Initiative (EW3), Union of Concerned Scientists, November 2011
- Water for Energy: Future Water Needs for Electricity in the Intermountain West, Pacific Institute, November 2011
- Effects of Climate Change on Federal Hydropower, Oak Ridge National Laboratory for DOE, draft July 2011, final forthcoming

generation capacity nationwide, are in counties associated with some type of water sustainability concern. The most significant future water stresses are in the South, Southwest, and Great Plains regions, with water use for electricity generation growing especially rapidly in the east (Figure 8), although water sustainability concerns are seen in many regions (Figure 9).

The report by the Union of Concerned Scientists starts with a number of cases where droughts and/or heat waves since 2006 have required reductions in electricity generation, with Texas as a current case in 2011 (Figure 10). It notes that droughts and heat waves are projected to be more frequent and more severe with climate change, which is a reason for concern not only in the U.S. west but also in a number of locations in the east (Figure 11). The report also notes that (a) water intensity varies regionally, along with water availability, and (b) low-carbon electricity technologies are not necessarily low-water in their input requirements. Finally, the report includes a host of ideas about how to reduce risks and threatsOne effort to compare operational water consumption for different sources of electricity is Figure 12 (SRREN, 2011).

Effects of rising ambient air and water temperatures

In addition to the problem of water availability, there are issues related to an increase in water temperature. Use of warmer water reduces the efficiency of thermal power plant cooling technologies. Also, warmer water discharged from power plants can alter species composition in aquatic ecosystems. Large coal and nuclear plants have, in several cases in recent history, been limited in their operations by reduced river levels impacting water intake structures, by higher temperatures, and by thermal limits on water discharge (UCS, 2011).

Situations where the development of new power plants is being slowed down or halted due to inadequate cooling water are becoming more frequent throughout the nation. For example, Cooley et al discuss several instances of reduced production, plant shutdowns, and revised configurations of proposed new plants driven by reduced water availability or anticipated constraints on new capacity. Current research at MIT for the DOE Regional Integrated Assessment Modeling (RIAM) project indicates that the key factor is EPA requirements that the water temperature in a power plant's "mixing zone" (where water emissions mix with ambient surface water) not exceed a standard related to impacts on river wildlife. Some regions, such as the Ohio River Basin, have multiple plants sharing the same water body and have regulatory constraints on the cumulative heat discharge, river temperature rise, and maximum river temperature.

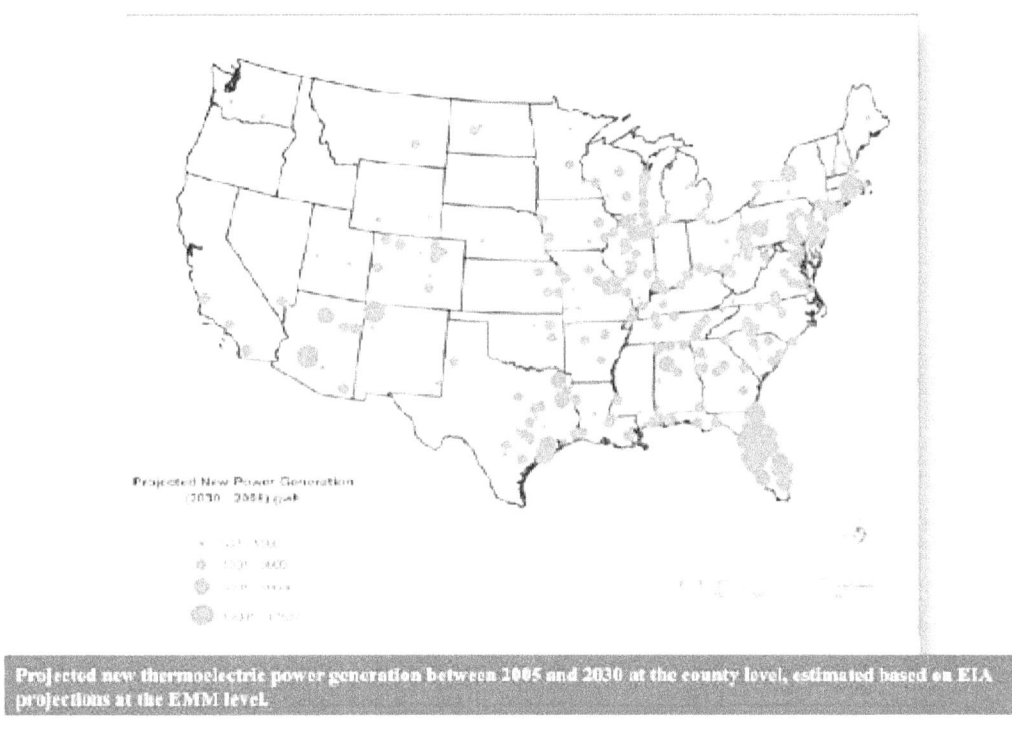

Projected new thermoelectric power generation between 2005 and 2030 at the county level, estimated based on EIA projections at the EMM level.

Figure 8. Water use for electricity generation and other sectors: Recent changes (1985-2005) and future projections (2005-2030), EPRI Technical Report, November 2011.

Historically, especially during seasonal droughts and/or heat waves, increases in ambient water temperature have sometimes required reductions in power output in order to avoid exceeding the EPA standard (i.e., to reduce warmer water discharges from the power plant). The alternative for many thermal power plant operators in the long run, if ambient temperature increases cannot be avoided, would be to invest in recirculating cooling systems, with high capital costs and some energy costs

.

The efficiency and output of thermal power plants, fossil or nuclear, is sensitive to ambient air temperatures as well; higher temperatures reduce power outputs.

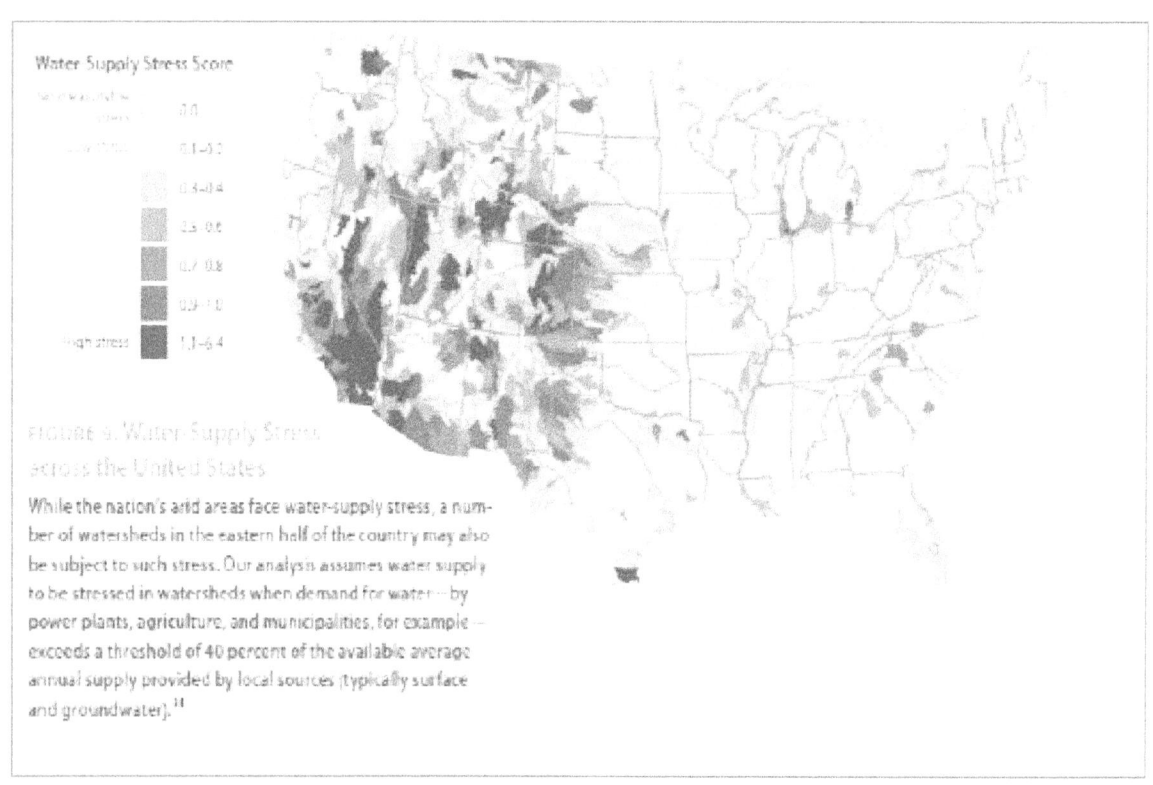

Figure 9. Water supply stresses due to demands for electricity generation and other sectors: Recent changes (1995-2005) and future projections (2050-2030), EPRI Technical Report, November 2011)

Steam cycles, which are used in most base load generation, are sensitive to cooling water temperature while combustion turbines used primarily for peaking generation are primarily sensitive to ambient air temperature. Gas turbines, which are dispatched primarily for daily and seasonal peaking service, are sensitive to ambient air temperatures. Figure 13 illustrates the effect of ambient temperature on the output and heat rate of a simple cycle combustion turbine.

Although these effects are not large in percentage terms, even a relatively small change could have significant implications for regional or national electric power supply. For example, an average reduction of 1 percent in electricity generated by thermal power plants nationwide would mean a loss of 33 billion kilowatt-hours per year, about the amount of electricity consumed by 3 million Americans, a loss that would need to be supplied in some other way or offset through measures that improve efficiency or reduce demand. This one-percent shortfall is roughly equivalent to the output of 5 GW of electricity generation capacity, operating at a typical capacity factor of 85%. The output falloff of combustion turbines at high temperatures can be particularly troublesome during high temperature events when peaking capacity is broadly dispatched to help meet electrical demand

Exposures to climate-related weather extremes and extreme events

A significant fraction of America's energy infrastructure is located in areas vulnerable to impacts of climate change, especially in coastal areas: power plants, oil refineries, facilities that receive oil and gas deliveries, and pipelines (SAP 4.5; GCRP 2009). Rising sea levels combined with more intensive coastal storms and, in the Gulf Coast land subsidence (SAP 4.7,) threaten direct losses, such as equipment damage from flooding or erosion, and indirect effects, such as the costs of raising vulnerable assets to higher levels or building new facilities farther inland, increasing transportation costs. As witnessed in 2005, hurricanes can have a debilitating impact on energy infrastructure. Direct losses to the energy industry in 2005 have

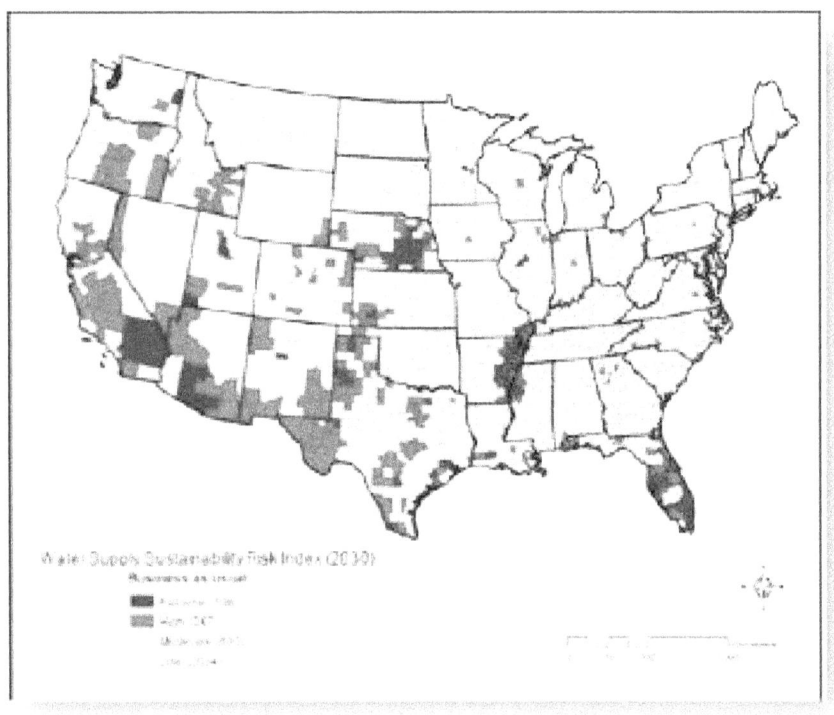

Figure 10. Water-supply stresses across the United States (Union of Concerned Scientists, 2011).

been estimated at $15 billion, with millions more in restoration and recovery costs. As one case, the Yscloskey Gas Processing Plant (located on the Louisiana coast) was forced to close for six months following Hurricane Katrina, resulting in lost revenues to the plant's owners and employees and higher prices for consumers, as gas had to be produced from other sources (SAP 4.5).

In fact, the National Oceanic and Atmospheric Administration warns that, outside of greater New Orleans, Hampton Roads is at the greatest risk from sea-level rise or any area its size. EPA (Titus, 2011) is exploring 'rolling easements' and other mechanisms of dealing with water intrusion in areas such as Hampton Roads, Virginia. The Hampton Roads and Norfolk area is home to significant energy facilities, including the Lamberts Point Coal Terminal, the largest on the East Coast, the Yorktown Refinery (now inactive), and the Dominion Yorktown power plant (~1200 MW) (Fears, 2011).

In nearby Chincoteague Island, VA, the Fish and Wildlife Service (2011) is evaluating management plan options that include expected losses of certain sea-facing areas and facilities. These are not, themselves, energy supply infrastructure but the planning is indicative of the high level of certainty regarding sea-level rise in the region. Moreover, many of California's power plants are at risk from sea-level rise, especially in the low-lying San Francisco Bay area (Figure 14).

But the impacts of an increase in severe weather are not limited to sea-level and hurricane-prone areas. For example, rail transportation lines, which carry approximately two-thirds of the coal to the nation's power plants, often follow riverbeds, especially in the Appalachian region. More intense rainstorms, which have been observed and projected, can lead to river flooding, which can "wash out" or degrade nearby rail beds and roadbeds. This is also a problem in the Midwest,;

Figure 11. Where power plants drive water supply stress (Union of Concerned Scientists, 2011).

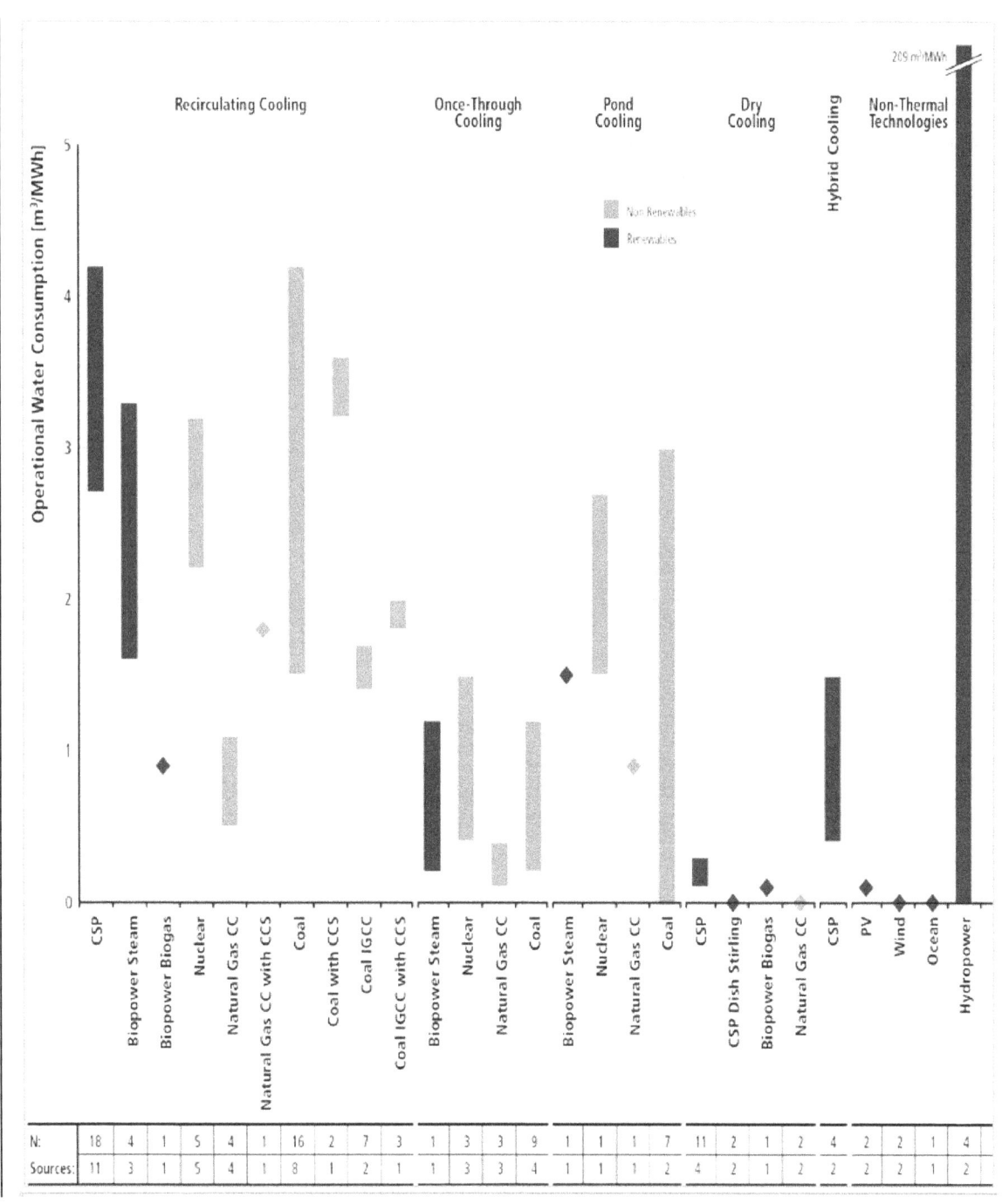

Figure 12. Ranges of rates of operational water consumption by thermal and non-thermal electricity-generating technologies based on a review of available literature (m3/MWh). Bars represent absolute ranges from available literature, diamonds single estimates; n represents the number of estimates reported in the sources. Note that upper values for hydropower result from few studies measuring gross evaporation values, and may not be representative (SRREN, 2011).

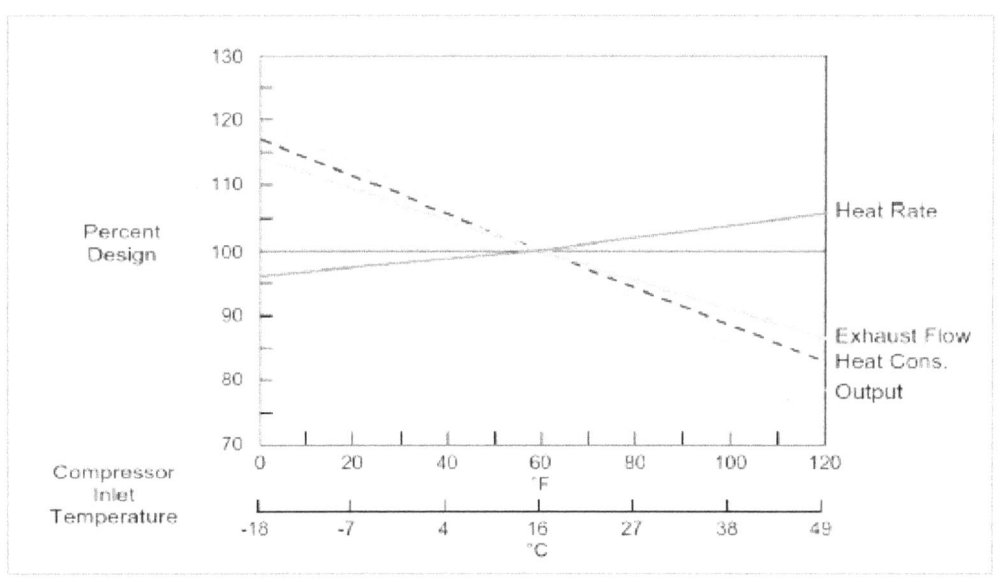

Figure 13. Effects of ambient temperature (GCRP 2009).

which experienced major flooding of the Mississippi River in 1993 and 2008 and is also vulnerable to climate change effects, from temperature changes to severe weather events.For instance, the year 2011 was marked by a February intrusion of severe cold weather into Texas, New Mexico, and Arizona which led to electrical blackouts and natural gas shutdowns (Souder, 2011, FERC, 2011); springtime flooding (Anada, 2011) in the Missouri and Mississippi River valleys; a prolonged heat wave and drought in the Southern Plains, particularly Texas (Burkhardt, 2011and a modest Hurricane Irene that tracked through the densely populated mid-Atlantic and Northeast regions, spawning many local power outages (Clayton, 2011).

Possible effects of climate change on electricity grid reliability have been studied by EPRI and NERC (EPRI and NERC, Joint Technical Summit on Reliability Impacts of Extreme Weather and Climate Change, 2008), and additional studies are being carried out by the California Energy Commission and others. The EPRI/NERC joint technical summit found that uncertainty is on the rise, calling for improvements in forecasts and a need to increase grid flexibility. Concerns include impacts of weather on patterns of demand on supply facilities such as wind power (affecting transmission demands), extreme summer power demands that can cause severe voltage depression, effects of higher temperatures on the lifetime of distribution transformers, and effects of high wind speeds on overhead power lines and risks from wildfires (Figure 15).

Overall, the nation's energy infrastructure is extensive, expensive, and diverse. Its size indicates that climate change impacts are unlikely to have a sizeable impact at the national scale, e.g., on the national Gross Domestic Product. Current information suggests that thermal electric power production will see modest reductions in output and efficiency, unless adaptations are undertaken, and that some

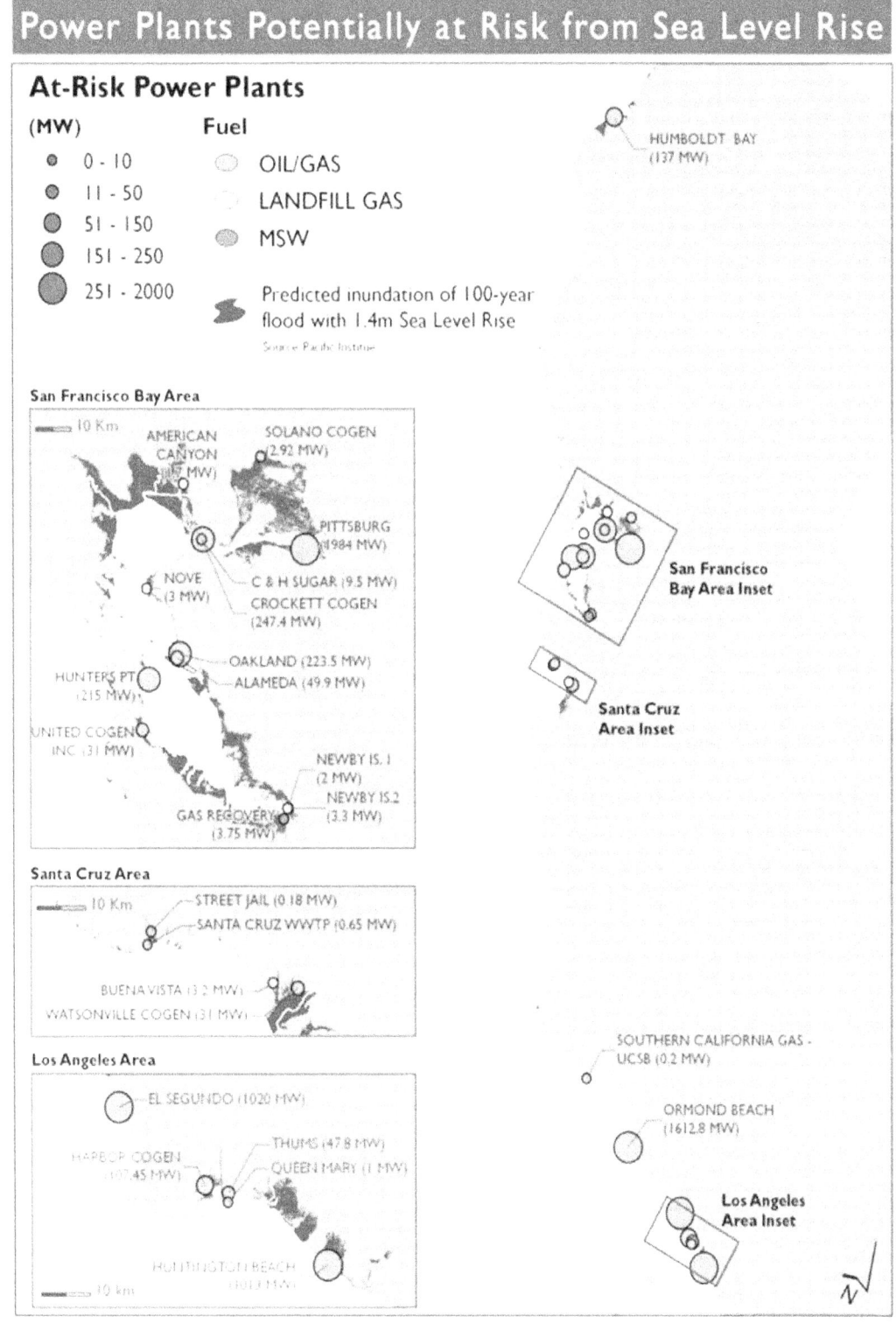

Figure 14. Power plants potentially at risk from sea level rise (Sathaye, et al., 2011).

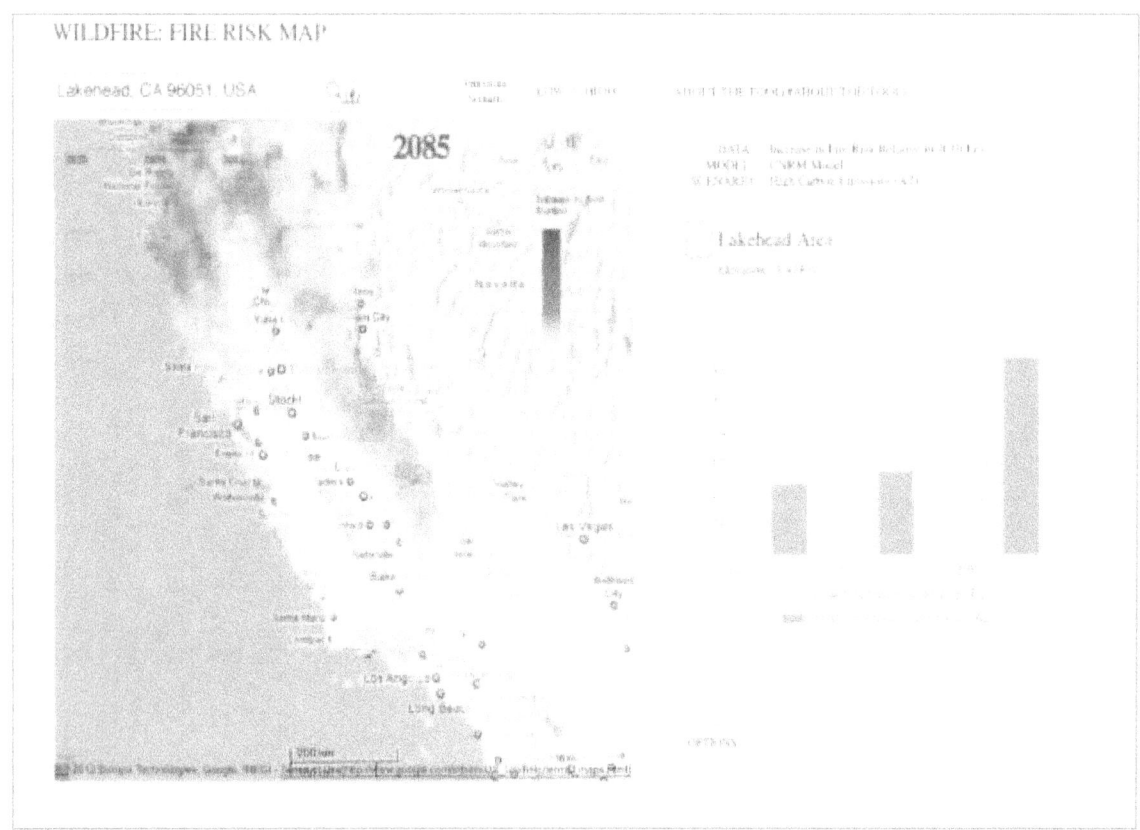

Figure 15. Projected wildfire risks in California, 2085 (Source: Cal-Adapt, 2012).

transmission and distribution capacities may also be reduced somewhat. The more serious issues are at a regional scale, often episodic over relatively short terms: e.g., floods, droughts/heat waves. A particular concern is coincident events: e.g., droughts combined with heat waves; and interdependencies are not fully storms, understood: e.g., between electricity and gas supply systems, or between regional electric utilities. Interregional and intraregional bulk power transmission has the potential to add resilience to the supply system, but such linkages can themselves be vulnerable to disruptions (as in the case of the 2003 Northeast blackout).

    3)   Renewable energy potentials.

Since SAP 4.5 and GCRP, 2009, knowledge about implications of climate change for renewable energy potentials has increased (see, for example, Table 4), although many answers await improvements in data. Toward that end, there has been substantial progress in understanding the need for, and initiating international collaborations to pursue, the downscaling of climate data under various emission scenarios to inform the assessment of renewable energy potentials. This includes efforts to both better reflect renewable resource potential within integrated models and to evaluate impacts of climate change on the future potential of these resources under various emission scenarios. While these developments reflect a much

34

Table 4. Recent Research on climate change implications for renewable energy

Topic Area	Sub Topic Area	Relevant Publications and Research
Renewable Energy Production	*General*	World Bank., 2011, SSREN, 2011,
	Biomass	Poudel, et al, 2011. Haber, et al, 2011. de Lucena, A. F. P., A. S. Szklo, R. Schaeffer, R. R. de Souza, B. S. M.C. Borba, I. V. L. da Costa, A. O. P. Júnior and S. H. F. da Cunha, 2009.
	Hydro	Hamlet, A. F., S. Y. Lee, K. E. B. Mickelson, and M. McGuire Elsner, 2009. de Lucena, A. F. P., A. S. Szklo, R. Schaeffer, R. R. de Souza, B. S. M.C. Borba, I. V. L. da Costa, A. O. P. Júnior and S. H. F. da Cunha, 2009 UPME, 2009. Hamlet, et al., 2009
	Ocean/ Hydrokinetic	World Bank., 2011, Harrison, G. P., and H. W. Whittington, 2002.
	Solar	Heath, G. A.; Burkhardt, J. J.; Turchi, C. S., 2011. Bard, E., and M. Frank, 2006. Kurtz, S.; Whitfield, K.; Miller, D.; Wohlgemuth, J.; Kempe, M.; Bosco, N.; Zgonena, T., 2009. GE Energy, 2010.
	Wind	Pryor, S. C., and R. J. Barthelmie, 2010. Sailor, D. J., M. Smith, and M. Hart, 2008. Bloom, A., V. Kotroni, and K. Lagouvardos, 2008. World Bank., 2011,
Energy Demand and Renewables		California Energy Commission (CEC). 1999. Sailor, David. 2001. Crowley, Christian and Joutz, Frederick. 2005. Sailor, D. and Ricardo Munoz. 1997. PNNL/DOE, 2008.
Renewable Energy and Water Nexus		Union of Concerned Scientists, November 2011. Pacific Institute, 2011. Scott, Christopher A., Suzanne A. Pierce, Martin J. Pasqualetti, Alice L. Jones, Burrell E. Montz, Joseph H. Hoover, 2011. Macknick, J.; Newmark, R.; Heath, G.; Hallett, K. C., 2011.
Integrated Analyses		Western Governors Association, 2010 Union of Concerned Scientists, November 2011. MacDonald, G. M., 2010. Ackerman and Stanton, 2011.

broader awareness of the need to understand potential climate impacts on renewable energy resources, the insights from these models, analyses, and existing case studies tend to provide information on the anticipated impacts on total generation from renewables but less insight on temporal impacts of these resources that effect energy supply (World Bank, 2011) . With this constraint, even improved downscaling of climate data and regional modeling will only marginally improve the understanding of unit and utility level generation impacts for renewables.

As a result, there still remains a need for more detailed spatial and temporal data on likely energy system impacts under various climate scenarios. Without these data to inform renewable energy supply estimates, planners have been responding by seeking to ensure more system flexibility (reservoir expansion and better management for hydropower development, alternative storage technologies for solar, improved transmission and dispatch protocols for wind, etc.) to manage these uncertainties. In addition, there have been significant improvements in accurate short-term forecasting for wind and solar resources over the last three to four years with commercial firms starting to fill this space by providing tools and forecasting data to utility clients.

The hydropower sector, because it is such a major component of many national and regional energy supply systems and serves as a primary base load resource for many countries, provides early examples of potential impacts of climate change variability on energy supply and production and strategies to address these risks (see Box 3: Implications of Climate Change for Hydropower Supply). Increased use of models to simulate river flow under different scenarios to assess hydropower generation in electricity generation (et al., 2009; de Lucena et al., 2009), along with the potential economic and financial implications for specific sites are being used on a more regular basis. These models rely on river flow series that are derived by downscaled GCM model data assessing precipitation and temperature under various emission scenarios again reflecting the need for better down scaled data sets (also see Box 4).

As noted in earlier assessments, it is anticipated that extreme events, air temperature, and atmospheric conditions will directly impact the efficiency, performance, and economics of all renewable energy technologies (SAP 4.5, 2008). At the same time, increased peak demand for cooling during the day and late afternoon may in some cases make utility scale PV and CSP more attractive and economic in particular regions. The body of research on these impacts continues to grow, generally indicating a variety of impacts – positive and negative – on potentials at a fine-grained (local) scale but very little impact on aggregate potentials at the national scale. In other words, reduced potentials in some areas are likely to be balanced by increased potentials in others; the main Impact on renewable energy potentials is likely to be a shift in national/regional patterns of potentials. But there is still a gap in the availability of very localized forecasts that can inform how specific sites or regions may be affected.

Box 3. Implications of Climate Change for Hydropower Supply

IPCC's Fourth Assessment Report and SAP 4.5 note that projected effects of climate change on regional rain and snowfall, both in terms of long-term changes in totals and changes in seasonal variability, are virtually certain to have implications for hydropower production in some US regions. Since 2008, most of the new research on energy/water connections has focused on consumptive uses of water by thermal power plants (see section III B) rather than water resource availability for hydropower, but ongoing research at ORNL for a draft report on federal hydropower (see Box 1 for reference), based on CMIP 3 ensembles of SRES scenarios, suggests several regional trends – varying seasonally, spatially, and temporally, with large uncertainty bounds.

In very general terms, this research indicates higher annual runoff in the US northwest to 2040, mainly in the spring, but a possibility of slightly decreasing hydropower generation in the longer term. It shows considerable variation across the west and southwest, with an overall slightly decreasing overall trend but with dry water years more frequent. In the southeast, it finds that dry years will occur significantly more often, while normal and wet years will decrease somewhat, associated with an overall slight decrease in hydropower generation but an increase in annual and seasonal variability in generation. The northern Great Plains region is the only US region projected to become wetter, with a potential to increase hydropower generation.

Over the past five years, industry, utilities, and governments have also received more practical exposure to the challenges that environmental changes can bring to energy security and reliability and the relative unpredictability of these impacts on renewable energy supply. These specific examples have provided more case studies and insights for the research community to better understand the potential 'relationship between climate change and hydropower and wind production in particular (SRREN, 2011: See Box 4). As the relative share of solar and dedicated biomass for energy production increases it is anticipated that these sectors will continue to demand better data and forecasting to inform project planning and financing.

4) Toward an integrated perspective.

Although it is customary to consider climate change implications by energy supply sector, many issues and options reach across sectoral boundaries and call for integrated modeling and analysis. Examples include bioenergy, which is rooted in renewable energy supplies but provides fuels to the liquid fuel industry and to thermal electricity generation, and renewable energy development and conversion;

Box 4. Special Report on Renewable Energy Sources and Climate Change Mitigation (SRREN)

In 2011, IPCC produced a 1075-page special Special Report on Renewable Energy Sources and Climate Change Mitigation (SRREN, 2011), intended to assess current knowledge about possible contributions of six renewable energy sources to the mitigation of climate change by reducing total greenhouse gas emissions: bioenergy, direct solar energy, geothermal energy, hydropower, ocean energy, and windpower. It notes growing energy demands worldwide, together with a rapid expansion in the use of renewable energy (RE) sources in recent years, and it projects that widespread growth will continue. It finds that global technical potentials for renewable energy are not a constraint on continued growth in RE use. Issues include higher levelized cost of RE from many sources than existing energy prices, although costs are generally declining; challenges of integrating some RE systems into current energy supply systems, related to such issues as scale, although integration is proceeding successfully in some cases; and supportive policy environments, complicated by the diversity of RE sources and systems. Finally, it notes that research to date on climate change impacts is very limited, the main current concern being about water availability for hydropower and some bioenergy systems.

and integrating mitigation and adaptation in energy strategy development. One important starting point for such integrated analysis is the set of perspectives and electricity supply; water issues, which cross boundaries between oil and gas use, tools of the integrated assessment research community, which is supported by DOE's Office of Science.

Sustainable trajectories for bioenergy development and use

Integrated assessments of sustainable bioenergy futures – affecting supplies for transportation fuels, electricity generation sources, and industrial and modern building heating -- have not yet considered changes in climate parameters extensively, although there has been some analysis of CO_2 concentration in the atmosphere as a factor affecting biomass productivity (SRREN, 2011). In general, bioenergy production potentials will be affected by both climate change and rising concentrations of atmospheric carbon dioxide. Increased concentrations tend to enhance crop yields, while changes in climate can either enhance or degrade yields, varying across regions and over time.

Some attention has been paid, however, to environmental variables that shape the sustainability of bioenergy development trajectories (Figure 16) and to frameworks for determining which variables are most important for sustainability (Figure 17).

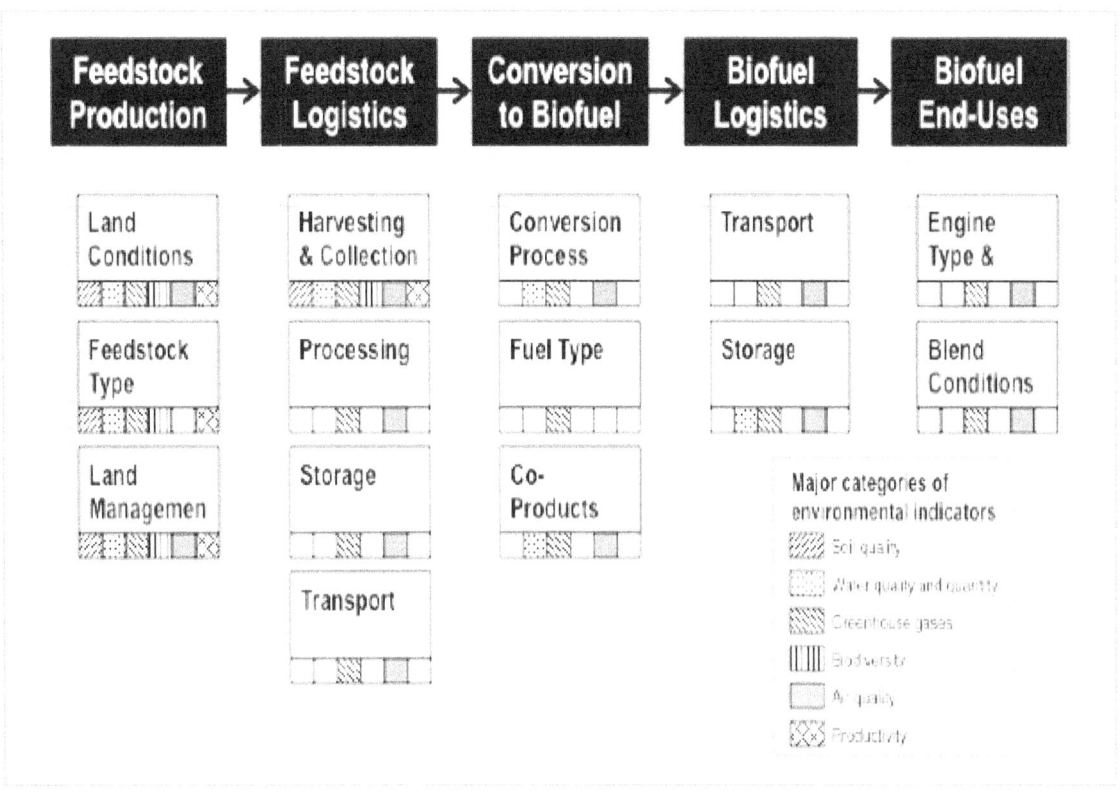

Figure 16. Contexts for environmental indicators of sustainability in the biofuel supply chain – DOE-EERE Office of Biomass Program - Sustainability

In many cases, the key variables appear to be land use and water availability; and research is under way to evaluate sensitivities to water availability, which can be subject to climate change. Initial analyses of climate feedbacks through bioenergy production and use systems have been carried out by PNNL, considering effects of climate change itself on biomass feedstocks such as corn and also effects of mitigation regimes on land use (Figures 18 and 19). In addition, some research has been carried out on effects of climate *variability* on bioenergy production, such as a study that found that maize production for ethanol production varies significantly with climate variability (i.e., ENSO phases) (Persson et al., 2009). Although most of the current attention is focused on ethanol from corn and other food-related crops, a growing emphasis is likely to be on lignocellulosic biofuels, such as switchgrass (especially for aviation and diesel fuels (EIA, 2010). An issue for integrated perspectives is how such new sources will fit into broader agricultural and forestry landscapes that are themselves being affected by climate change.

Integrating Energy, Water, and Climate in the American West

An integrated regional approach to evaluating energy, environmental, and land use factors for policy planning has gained prominence in recent years. Whether

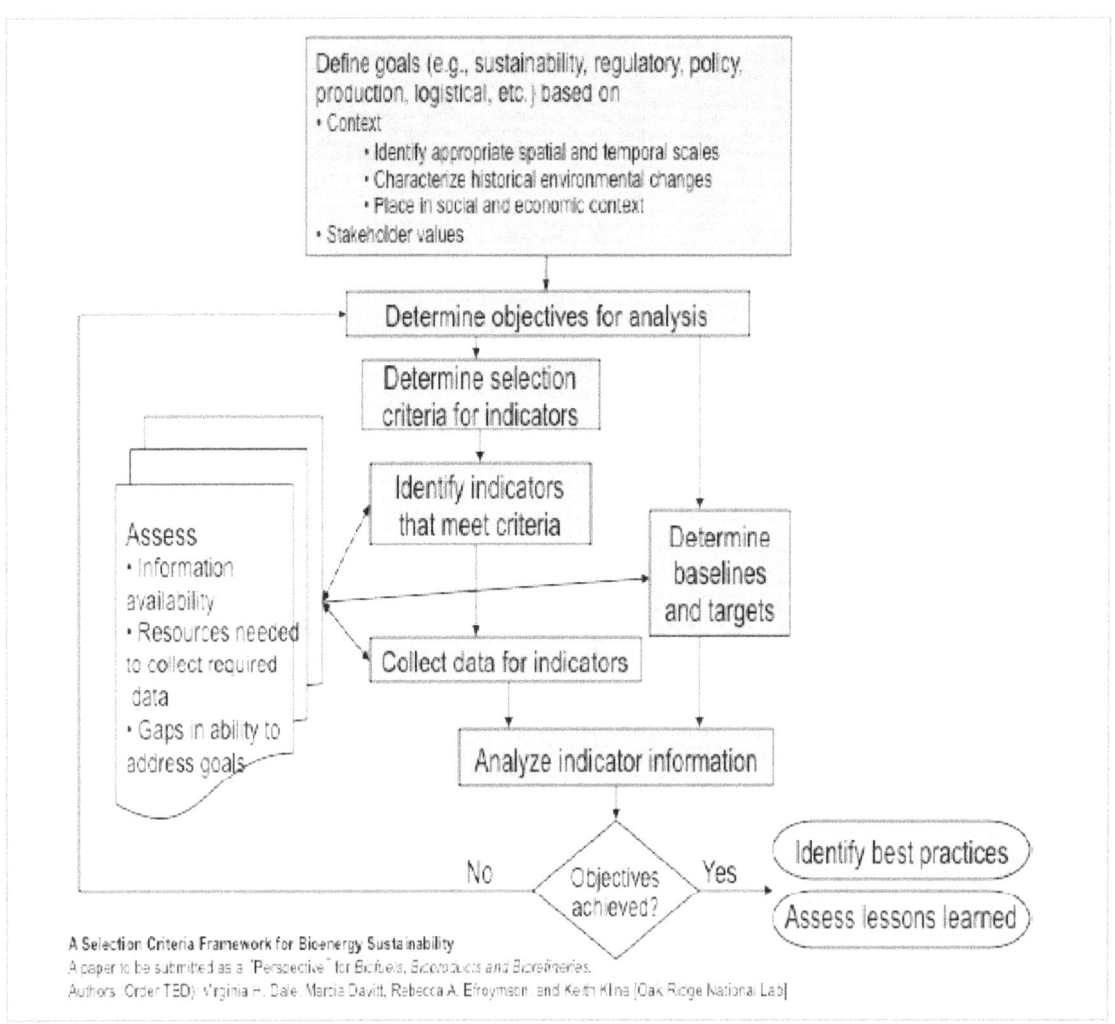

Define goals (e.g., sustainability, regulatory, policy, production, logistical, etc.) based on
• Context
    • Identify appropriate spatial and temporal scales
    • Characterize historical environmental changes
    • Place in social and economic context
• Stakeholder values

Determine objectives for analysis

Determine selection criteria for indicators

Identify indicators that meet criteria

Assess
• Information availability
• Resources needed to collect required data
• Gaps in ability to address goals

Determine baselines and targets

Collect data for indicators

Analyze indicator information

No   Objectives achieved?   Yes

Identify best practices

Assess lessons learned

A Selection Criteria Framework for Bioenergy Sustainability
A paper to be submitted as a "Perspective" for Biofuels, Bioproducts and Biorefineries.
Authors (Order TBD): Virginia H. Dale, Marcia Davitt, Rebecca A. Efroymson, and Keith Kline [Oak Ridge National Lab]

Figure 17. Framework for Selecting Sustainability indicators for Bioenergy

organized by watershed, utility service area, or grid, these analyses seek to answer questions of energy security, sustainability, and operational optimization that cannot be addressed within traditional geographic or political boundaries. Water often plays a unifying role in this research, especially in the American West, as policymakers seek to better understand the implications of one of the most direct and politically sensitive aspects of climate change on energy and land use planning.

Although recent analyses have examined linkages in a number of regions (e.g., section III b above), the US West has been the focus of particular attention. This is due in part to the practical realities of population growth, increased energy and water demand, and existing vulnerabilities to environmental change but also to the political leadership of groups like the Western Governors Association that are advancing aggressive renewable energy strategies and recognize the vulnerability of their local economies to climate change. This region is also historically drought prone with multiple local, State, federal regulations impacting the use of water across competing demands including agriculture, industry, energy, and residential.

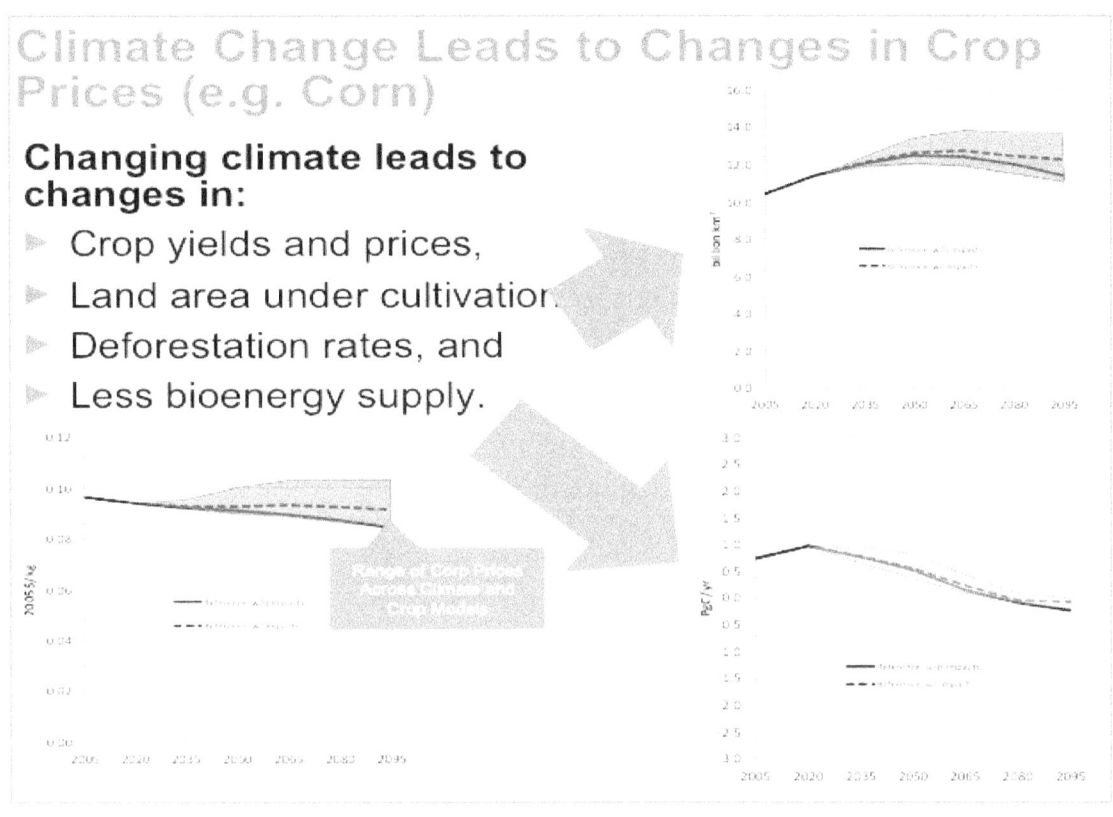

Figure 18. Climate change leads to changes in crop prices (e.g., corn) (Calvin, K., et al., 2012).

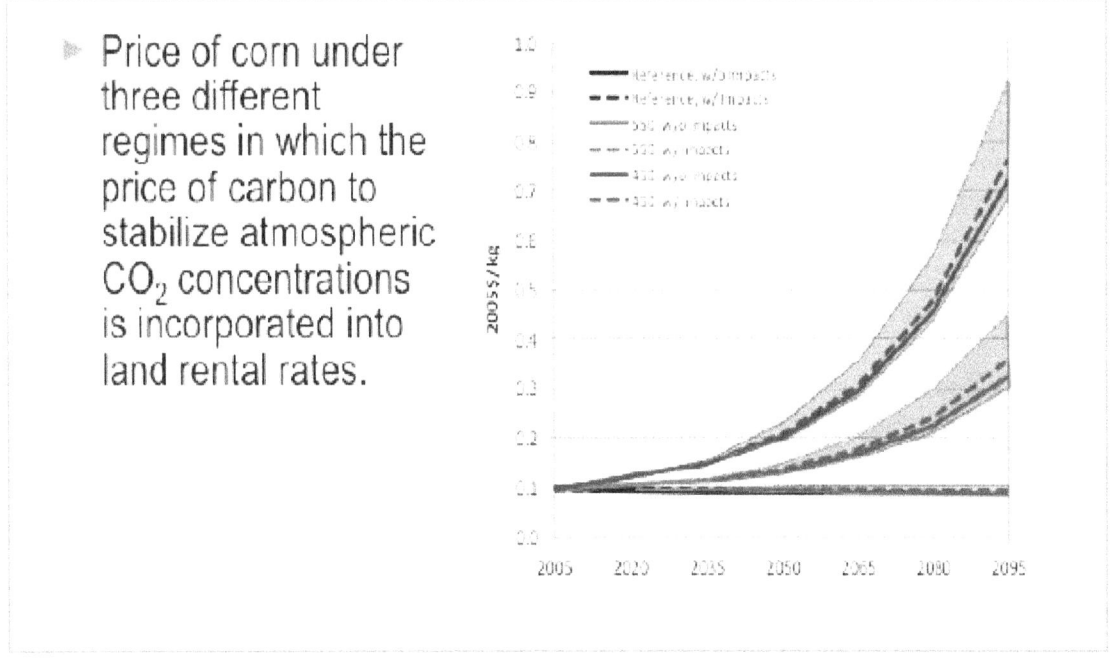

Figure 19. Mitigation regimes also affect crop prices and land use (Calvin, K., et al., 2012).

use. With the energy sector placing increased demands on this limited resource, an improved understanding of current water requirements in the energy sector is critical for near term decision making on permitting of projects as well as longer term capacity expansion planning. When climate change considerations are incorporated, this calculus becomes more complicated, both in terms of the trade-offs across specific power technologies and their relative exposure to future changes in temperature or water availability. Further, both water and land use considerations influence decision making in Western States that often rely on an agriculture base for their economy but are also endowed with significant renewable and traditional energy resources to exploit. Research, analysis, and data that can help inform how resource and water availability may change under different climate change scenarios is therefore of high value to State and regional planners. Even where significant uncertainties exist, the ability to better forecast and plan for possible scenarios will help ensure a more flexible, and timely policy response to environmental stresses and to understand the integrated nature of water, land use, and energy in that region.

For the renewable energy sector, for instance, improved forecasting and analysis of climate change impacts is critical. Transmission planning at a regional level and state approvals for capacity expansion will directly influence the feasibility and cost of large states like California in meeting renewable portfolio standard (RPS) targets that will be dependent on energy imports from neighboring states. If water or other environmental constraints affect energy production or the approval of renewable energy projects, the long-term cost of Western states in meeting their RPS goals will increase significantly. As states also consider alternative fuel and transportation strategies, a nexus of land use, water, and energy will also emerge as policy makers consider the optimization of land and biomass resources for power, fuel, or food under increasing variable, and uncertain, hydrologic cycles in future. Ensuring that the quality of data, analysis, and accurate scenario development to inform these integrated assessments keeps pace with policy design and implementation will be critical, particularly in the West and other regions that may be most vulnerable to climate change.

Recent research taking an integrated approach to addressing energy, water, and land-use considerations is summarized and assessed in the NCA Technical Report on Water/Energy/Land System Interactions; examples include Pacific Institute, 2011; UCS, 2011; EPRI, 2011, MacDonald, 2010; Ackerman and Stanton, 2011; Kenny and Wilkinson, 2012; Scott et al., 2011; and Western Governors Association, 2010.

Integrating mitigation and adaptation in energy strategy development.

The challenge of integrating climate change adaptation and mitigation has received some recent international research attention (e.g., Wilbanks et al., 2007; Wilbanks and Sathaye, 2007; Ayers and Huq, 2009; and NRC, 2010 and 2011) including an analysis of relationships between nuclear power siting, as an aspect of decarbonizing electricity production, and adaptation to sea-level rise (Kopytko and Perkins, 2011). This challenge was the focus of a chapter of IPCC's Fourth Assessment Report from Working Group II (Impacts, Adaptation, and Vulnerability) and is also a focus of the emerging Fifth Assessment Report. An example of perspectives arising from this work is Table 7 from the NAS/NRC report on America's Climate Choices.

Among the topics that have received attention are sustainable bioenergy development, especially if feedstock choices move toward wood and grass sources rather than crops; locational choice for energy supply facilities related to areas vulnerable to impacts of climate change; improvements in the efficiency and affordability of air conditioning in residential and occupational buildings as ways to extend space conditioning benefits to a larger share of the population in a warming world without significantly increasing carbon emissions from electricity generation; and enhanced regional connections as ways to add flexibility to risk management strategies surrounded by uncertainties about future conditions.

A further topic that could emerge is geo-engineering as a climate change response option related to both mitigation and adaptation (see NRC 2010 and 2011).

5) Indirect impacts of climate change on energy systems.

SAP 4.5 broke new ground partly by recognizing that impacts of climate change on energy systems are related not only to direct impacts, such as reduced snowfall on hydropower potentials, but also to indirect impacts. Examples cited in that report included possible effects on energy planning, energy technology development and use, energy institutions (and supporting institutions such as finance and insurance), energy-related dimensions of regional and national economies, energy prices, environmental emissions, energy security and energy technology and service exports. The report also noted that climate change effects in other countries could affect US energy supply and use.

In the period since 2007, some of the issues have not received significant further research attention – such as implications for regional economies of changes in energy resource/technology trajectories – but several issues have been examined in further detail.

a) Relating climate change responses and energy security concerns:

Relationships between national energy strategies and U.S. energy security have been a topic of discussion since the 1970s. In recent years, this issue has been connected directly with climate change. For instance, a "Climate Change War Game" was organized by the Center for New American Security in July 2008 to explore national security implications of global climate change (http://www.cnas.org/node/956). Most recently, Faeth, 2012, has raised questions about water requirements for a number of energy options related to U.S. energy

Table 5. Matrix of interdependencies among the different elements of a national response to climate change (NAS, 2011)

	Will strengthen this element because …			
	Limiting	Adapting	Advancing Science and Technology	Informing
Limiting		There may be less stringent, disruptive requirements (and thus lower costs) for adapting to climate change impacts	There may be less pressure to develop risky and/or expensive technologies for coping with impacts.	The decision environment may be less contentious if the severity of climate change can be limited
Adapting	Any given degree of climate change may be associated with less severe impacts and disruptions of human and natural systems		There may be less pressure to develop risky and/or expensive technologies for limiting climate change some forms of geo-engineering	The decision environment may be less contention if communities and key sectors are prepared to deal with impacts.
Advancing Science and Technology	R&D could help identify more and better options for limiting climate change	R&D could help provide more adaptation options and more knowledge about their implications		The knowledge base for informing decisions may be more complete, and the knowledge base about how to most effectively inform may allow better information flow
Informing	Effective options for limiting climate change may be more widely deployed and used	Effective options for adapting to climate change may be more widely deployed and used	Science may be more attuned to decision needs, and public support for advances in science is likely to increase	

security. Meanwhile, a 2010 paper in *Energy Policy* (Greene et al.) examined requirements for technological progress in eleven technology areas in order to achieve both CO_2 emission reduction and reduced oil dependence, concluding that each technology area must have a much better than 50/50 probability of success and that five technology areas (such as carbon capture and sequestration) are virtually essential.

More specifically, environmental dimensions of energy security have been examined by Brown and Dworkin (2011) and Brown and Sovacool (2011), who note that global financial markets – with which the US energy sector is linked – are subject to climate change vulnerabilities in many parts of the world. A recent reminder of possible vulnerabilities of supply-chain linkages as well as financial linkages has been the effect of the Fukushima nuclear disaster in Japan on supplies of electronics, generators, and turbines for electricity construction projects. Flooding and other extreme weather events can affect areas to which manufacturing has been outsourced, adding to energy security concerns, at least in the short term.

b) Technology research and development to expand the range of response options:

Another issue that has been discussed actively at the annual Energy Modeling Forum and is also addressed by NAS, 2011, is the role of technology research and development in making the stabilization of greenhouse gases in the atmosphere more feasible and affordable. For example, a special issue of *Energy Economics* in 2011 examined in some detail the economics of technology development and deployment to combat climate change. The issue stressed the necessity of rapid technological change if the rate of climate change is to be moderated.

c) Effects of climate change responses on energy prices

Effects of climate change responses on energy prices, especially climate policies to promote greenhouse gas emission reduction, have been a focus of considerable debate and associated analysis. A major stimulus has been the series of proposals for climate policy legislation before the U.S. Congress, such as the Low Carbon Economy Act of 2007 (Bingaman-Specter), the Climate Stewardship Act of 2008 (Lieberman-Warner), the Clean Energy Jobs and American Power Act of 2009, the American Clean Energy and Security Act of 2009 (Waxman-Markey), and the American Power Act of 2010, where impacts on electricity prices are a leading issue: e.g., (CBO, 2009). Proposed actions by state governments have also generated economic impact analyses, as have discussions of such energy technology options as carbon capture and storage and such energy policy options as renewable energy portfolio standards (e.g., NRC, 2009 and 2010). Also see the NRC, 2009, report on *Hidden Costs of Energy: Unpriced Consequences of Energy Production and Use.* One continuing theme is that, for the longer term, price effects of energy efficiency improvements and energy supply technology shifts depend considerably on success with technological innovation – a theme that dates back to DOE laboratory studies in the late 1990s.

C.   *Assessment Findings*

Regarding implications for components of energy supply systems and cross-cutting implications for energy supply and use, we find that:

Implications for components of the nation's energy supply and use systems

- **In most cases, the major current risk for both supply and use is from episodic disruptions related to extreme weather events**

High consensus, strong evidence

> **See Section III B, 1, 2**

- **Impacts from weather phenomena associated with climate change pose risks of economic costs to energy suppliers and users**

High consensus, moderate evidence

> **See Section III A, B**

- **Increases in average temperatures and temperature extremes will mean increasing demand for electricity for cooling in every US region, along with reductions in energy demands for space heating**
-

High consensus, strong evidence

> **See Section III A 1, 2, 8**

- **Climate change is expected to have a larger impact on peak electricity demand than on monthly average demand**

Moderate consensus, some evidence

> **See Section III A 2, 8**

- **Impacts of climate change are risks to many oil and gas supply activities in vulnerable coastal areas, offshore production areas, and tundra areas**

High consensus, moderate evidence

> **See Section III B, 1**

- **Both climate change and rising concentrations of atmospheric carbon dioxide will affect bioenergy production potentials**

High consensus, strong evidence

> **See Section III B, 3**

- **Expected seasonal and/or chronic water scarcity represent risks of electricity supply disruptions in many US regions**

High consensus, strong evidence

> **See Section III B 2**

- **Climate change will affect the geographical pattern of renewable energy supply potentials in the US**

Medium high consensus, moderate evidence

> **See Section III B 3**

- **Expected reductions in precipitation in the form of snowfall in the US West will reduce hydropower production, at least in some parts of the region**

High consensus, strong evidence

> **See Section III B 3**

- **In most cases, adaptation measures can reduce risks and prospects of negative consequences for energy supply and use**

High consensus, moderate evidence

> **See Section III A, B, IV**

Cross-cutting implications for energy supply and demand

- **Energy system resilience will benefit from progress with technology R&D**

High consensus, moderate evidence

> **See Section III A, B, IV C**

- **Most vulnerabilities and risks for energy supply and demand reflect relatively fine-grained place-based differences in situations**

High consensus, strong evidence

> **See Section III A, B**

- **The variability of risks from weather-related events in both time and space will increase with climate change**

High consensus, moderate evidence

> **See Section III A, B**

- **Climate change implications interact with and are affected by regulatory environments**

High consensus, strong evidence

> **See Sections III A, B**

- **In many cases, gaps in the availability of data limit the capacity to answer key assessment questions**

- *High consensus, strong evidence*

<table>
<tr><td>See Section III B; IV B</td></tr>
</table>

IV.   Implications for Future Risk Management Strategies

GCRP, 2009, notes that the US energy sector is large and complex, with impressive financial and management resources, capable of responding to major challenges. It is accustomed to strategy development and operation in the face of uncertainties and risks, both environmental and political. No sector has better capabilities to respond to challenges posed by climate change impacts.

In responding to the need to assure resilience in the face of such challenges, every credible source indicates that the appropriate strategy for energy supply and use is rooted in risk management for an uncertain future rather than precise impact projections for optimal decisions – not only seeking to reduce vulnerabilities but also to identify market opportunities.

For energy supply and use, strategies for managing risks associated with climate change will vary by resource/technology trajectory, institution, and climate change impact threat. Examples of adaptation measures that could be considered are summarized in Table 8, drawn largely from the report on adapting to impacts of climate change that was part of the NAS/NRC America's Climate Choices Report (NRC, 2010; World Bank, 2011). The World Bank report also includes a number of examples of climate change adaptations being implemented by the energy sector in other countries

In reviewing current knowledge about these and other possible adaptation options, some common elements of energy sector strategies can be suggested (NRC, 2010 and 2011; SAP 4.5; Bierbaum et al., 2007; IPCC, 2007; SREX, 2011): elements of risk management strategies, approaches likely to be taken, tools to help get the job done, and issues that should be considered.

A.   *Management Strategies*

Risk management is a major theme throughout NCA 2013, examined in detail elsewhere (also see NRC, 2011). For energy supply and use, it includes the following commitments by all major parties, public and private:

- Monitoring, evaluating, and learning from emerging experience with impacts and responses. Given extensive uncertainties about climate change impacts at particular times and in particular places and about payoffs of specific adaptation strategies, it is important to observe, evaluate, and reconsider risks and responses iteratively, sharing lessons learned as appropriate.

- Increasing flexibility in order to manage uncertainties: e.g., regarding population trends, emerging impacts, and policy environments. Given uncertainties about not only climate change itself but also about future trends in socioeconomic and policy conditions, it is important to stress flexibility – rooted in a continuing learning process – in order to assure an ability to handle unexpected developments and surprises.

- Reducing system sensitivities: e.g., to water scarcity, temperature increases, exposures in vulnerable areas. Where energy supply and use systems are especially sensitive to climate-related parameters that are likely to be sources of stress, risk management will include attention to ways to reduce those sensitivities through changes in technologies, materials, and corporate strategies.

- Focusing on adaptation opportunities provided by structures or equipment that are toward the end of their lifetimes (or performing poorly) so that changes are required. Energy supply and use systems are built on structures and equipment with finite lifetimes, and in any given year many such physical items are due for replacement. Decisions at those times are opportunities to move systems in directions that are better-adapted to climate change risks, usually at a lower net cost than retrofitting structures and equipment that will continue to be used for some time.

    Encouraging incentive structures that promote innovation. Risk management strategies nearly always benefit from innovation. Because innovation usually carries with it some degree of risk, since the new approach has not been fully validated by experience, it tends to emerge more quickly when it is supported by incentives – within, and especially external to, energy institutions. In many cases, this can be a fertile area for public-private sector cooperation in the national interest.

- Identifying strategies that offer prospects of net value rather than net cost ("value chains"). As suggested in section III B 1, risk management will be more aggressively pursued if it is imbedded in actions that offer value added, not just costs avoided. In many circumstances, especially if and as market conditions are "greening," this is a case that can be made

Table 6. Examples of adaptation measures to reduce losses/risks in energy systems World Bank, 2011

Energy System	TECHNOLOGICAL			BEHAVIORAL	
	"Hard" (structural)	"Soft" (technology and design)	Re(location)	Anticipation	Operation and Maintenance
SUPPLY — **Mined Resources** (incl. oil & gas, thermal power, nuclear power)	Improve robustness of installations to withstand storms (offshore), and flooding/drought (inland)	Replace water cooling systems with air cooling, dry cooling, or recirculating systems Improve design of gas turbines (inlet guide vanes, inlet air togging, inlet air filters, compressor blade washing techniques, etc.) Expand strategic petroleum reserves Consider underground transfers and transport structures	(Re)locate in areas with lower risk of flooding/drought (re)locate to safer areas, build dikes to contain flooding, reinforce walls and roofs	Emergency planning	Manage on-site drainage and runoff Changes in coal handling due to increased moisture content Adapt regulations so that a higher discharge temperature is allowed Consider water re-use and integration technologies at refineries
Hydropower	Build de-siting gates Increase dam height Construct small dams in the upper basins	Changes in water reserves and reservoir management	(Re)locate based on changes in flow regime		Adapt plant operations to change in river flow patterns Operational complementarities with other sources (for example, natural gas
Wind		Improve design of turbines to withstand higher wind speeds	(Re)locate based on expected changes in wind speeds (Re)locate based on anticipated sea level rise and changes in river flooding		

51

Table 6. Examples of adaptation measures to reduce Losses/risks in energy systems (continued)

Energy System	TECHNOLOGICAL		BEHAVIORAL		
	"Hard" (structural)	"Soft" (technology and design)	Re(location)	Anticipation	Operation and Maintenance
Solar		Improve design of panels to withstand storms	(Re)locate)based on expected changes in cloud cover	Repair plans to ensure functioning of distributed solar systems after extreme events	
Biomass	Build dikes Improve drainage Expand/improve irrigation systems Improve robustness of energy plants to withstand storms and flooding	Introduce new crops with higher heat and water stress tolerance Substitute fuel sources	(Re)locate based in areas with lower risk of flooding/storms	Early warning systems (temperature and rainfall) Support for emergency harvesting of biomass	Adjust crop management and rotation Adjust planting and harvesting dates Introduce soil moisture conservation practices
Demand	Invest in high-efficiency infrastructures and equipment Invest in decentralized power generation such as rooftop PV generators or household geothermal units		Efficient use of energy through good operating practice		
Transmission and Distribution	Improve robustness of pipelines and other transmission and distribution infrastructure Burying or cable re-rating of the power grid		Emergency planning	Regular inspection of vulnerable infrastructure such as wooden utility poles	

B.   *Approaches that support risk management*

The references listed above, along with discussions of risk management elsewhere in the NCA process and the energy supply and use technical input workshop discussion, suggest a number of approaches that are often useful, including the following.

- Vulnerability assessments. The starting point for any risk management strategy is a vulnerability assessment (NRC, 2010; also see section III B 1), which considers possible exposures to risk under a range of possible future trends and conditions.

- Partnerships. Risk management benefits from risk-sharing, e.g., through insurance coverage; but it also benefits from other kinds of partnerships as well. Because risks are embedded in such a wide variety of drivers and stresses, and no one institution is the best at assessing all of them, there are benefits to maintaining partnerships that enable information sharing about risks, response strategies, and emerging experience and lessons learned. Partnerships also help to identify actions being taken in other sectors and sub-sectors that could have consequences to an energy supplier or user, and they reduce pressure on particular institutions to play roles that are better played by others.

- Innovation, including regulatory structures that promote innovation and resilience. In almost every case, there are alternatives for reducing risks that are based on going beyond currently available technologies and practices. Pursuing, developing, and deploying innovative approaches can often reduce the net cost and increase potentials for implementing risk management strategies (see section IV C below). Although legal, regulation, or policy impediments to innovation may need to be addressed in some cases

- Bundling climate change responses with other agendas. When climate change risk management can be associated with risk management, stress reduction, and resilience enhancement in other connections as well – such as revitalizing infrastructure, reducing energy costs, multi-stress emergency preparedness, and/or reducing regional environmental impacts – then it is virtually certain to attract more widespread buy-in.

- Global linkages and risk management. Risk management for energy supply and use is related to the larger global context in both directions. For example, it can be connected with international linkages between energy systems (e.g., water from the Colorado River basin), and it can benefit from information on risks and responses in other countries (NRC,

2010). At the same time, risk management approaches in the US, including technology and policy innovations, can represent opportunities in global markets that are in some cases responding to climate change implications at least as actively as the US.

- Global technology transfer and cooperation. Responses to energy sector vulnerabilities and risk associated with climate change will be not only supported but in some cases enabled by developments in the global energy technology marketplace. For example, technology developments and experiences with technology applications for risk management in other countries may be useful in considering US menus for action, and much larger global markets for innovative technologies and practices may encourage innovation by US private and public sectors.

C.   *Tools That Will Be Useful*

The energy supply and use workshop, which stressed brainstorming about risk management approaches, identified two kinds of tools that would appear to be highly useful for risk management, both rooted in energy sector innovations that are under way for other reasons as well.

- Targeted technological change, e.g., for electricity generation peak-shaving or reduced water consumption. Where climate change introduces risks to energy supply and use, agendas for technology research and development can include risk reduction as a priority. For example, Figure 20, drawn from World Bank 2011, indicates how emerging technologies can reduce the water intensity of electricity supply.

- Smart systems. One of the frontiers of energy supply and use research is the increased use of information technology applications such as "smart grid" and sensors to enable monitoring and control feedback, increasing efficiency and flexibility and substituting intelligence for resource and materials consumption.

Once again, tools such as these represent opportunities to combine climate change risk reduction with technology innovation and modernization in the US energy sector in ways that offer multiple co-benefits.

D.   *Issues To Be Resolved*

Finally, both available source materials and the workshop discussion point to several cross-cutting issues for risk management strategies by and for the US energy sector.

.

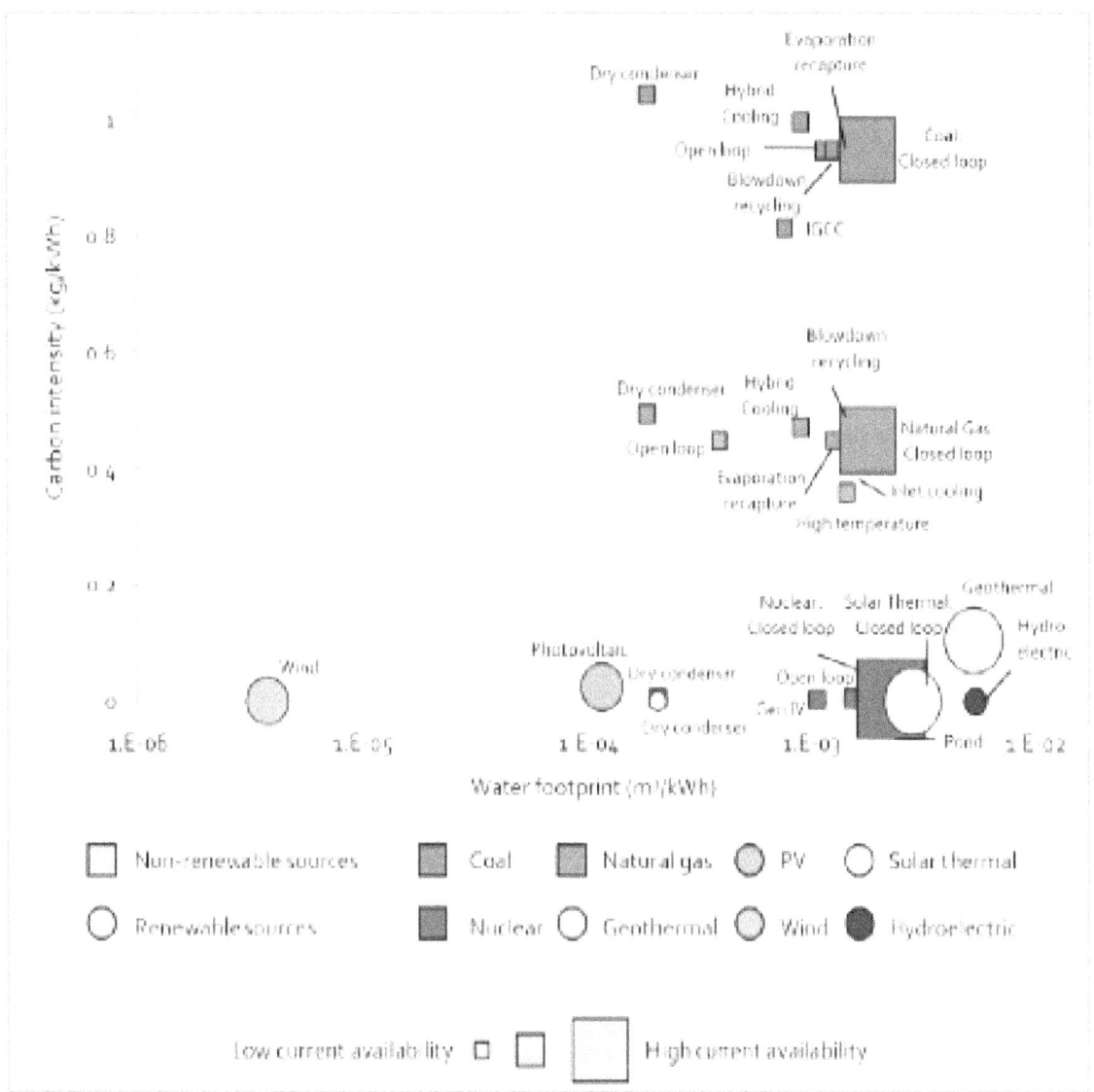

Figure 20. Effects of emerging technologies on carbon and water intensity of electricity sources (Source: Lux Research, 2009).

- Adapting to extremes. An assessment finding in section III D of this report notes that the major current risks of disruptions to energy supply related to climate change are from extreme weather events, and longer-term risks are associated with changes in climate and weather extremes, such as droughts, heat waves, and significant sea-level rise. The IPCC special report on

- Managing the Risks of Extreme Events and Disasters to Advance Climate Change Adaptation (SREX, 2011) responds to perceptions of decision-makers that adapting to risks of extremes and extreme events is different, and often more difficult, than adapting to gradual changes – especially when many disruptions are high-consequence/low-probability events for particular local contexts but high-probability for larger regions.

- Locational strategies relative to especially vulnerable areas. As noted above in this report, many risks to energy supply and use in the US are associated with locations in areas especially vulnerable to such projected climate change impacts as (a) more intense coastal storms together with sea-level rise and (b) significant chronic or seasonal water scarcity. In many cases, near-term strategies involve incremental buffering or hardening of existing systems in place, but such sources as SAP 4.7 (2008) – which projects apparent sea-level rise in the Gulf Coast area of 2-4 feet by 2050 – suggest that longer-term risk reduction strategies may have to consider the relocation of infrastructures and jobs (e.g., Kates, Travis, and Wilbanks, forthcoming). Because such contingencies involve complex relationships with regions and localities and their citizens and policymakers, they raise obvious issues for stakeholder interactions as well as corporate planning.

- Relationships between climate change mitigation and adaptation. The energy sector, especially on the supply side, is both the global and the national focus of climate change mitigation: efforts to reduce greenhouse gas emissions, particularly carbon dioxide. As a result, there is usually no way to avoid mitigation issues in considering impact risk-reducing adaptation strategies. Such relationships are notably salient in considering resource/technology portfolios for electricity generation, affordable renewable energy strategies, potential roles of natural gas in energy supply and use, and efficiency improvement in energy end uses associated with carbon emissions, such as transportation. A notable example is bioenergy production, where major production would have land use and carbon uptake implications as well as alternative fuel implications.

  As indicated in section III B 4, analyses of possible synergies between mitigation and adaptation strategies for energy supply and use are scarce at this point, but it appears very likely that there are more opportunities than have been explored to date between efficiency and redundancy. Literatures on emergency preparedness and community resilience (e.g., Cutter, et al., 2008) tend to stress the importance of redundancy as a way to increase the capacity to cope with surprises. Examples at a community or regional scale include stockpiling of critical supplies, such as electricity generators. Examples at a corporate scale include fuel reserves as an example at the national scale is the US Strategic Petroleum Reserve. But redundancy is not without costs; and where the benefits of such backups are more widely spread than the benefits of efficiency during normal operations, issues may emerge about how much redundancy is desirable and who pays for it.

- Sensitivity to what citizens/consumers/stakeholders want. In a democratic society such as the US, a risk management strategy that provokes public opposition is likely to be difficult to implement, regardless whether it appears to be good business. Conversely, a risk management strategy that generates broad public support is likely to have positive benefits in terms of the reputation of the energy institutions involved and even the potential for policy support. How to assure such sensitivity during the process of developing risk management strategies differs across sub-sectors of the US energy supply and use system, but it is an issue that always merits attention.

E.   *Assessment Findings*

Regarding climate change risk management strategies for energy supply and use, we find that:

- **Despite uncertainties about climate change impacts in the future, robust risk management strategies can be developed and – in an iterative manner that incorporates continuing observation, evaluation, and learning - implemented**

High consensus, moderate evidence

> **See Sections IV A, III A 8**

- **Many of the elements of such strategies can be identified based on existing knowledge**

High consensus, moderate evidence

> **See Section IV A**

- **A critically important step toward developing such strategies is conducting vulnerability assessments**

- *High consensus, moderate evidence*

> **See Section IV A 2**

V. Knowledge, Uncertainties, And Research Gaps

In a sector that has not been the focus of significant climate change risk and impact research, the knowledge gaps and uncertainties are profound and virtually unbounded. But participants in the preparation of this study, including the expert workshop, and several previous assessments (SAP 4.5; IPCC, 2007; NRC, 2010; Wilbanks, 2010), suggest that the following needs are at least illustrative of a number of high priorities for research, tool development, and learning.

A. *The Landscape of Needs for Knowledge*

In order to support assessments with a high level of confidence of implications of climate change for energy supply and use, there is broad agreement about the kinds of knowledge that are needed: knowledge to support response actions, fundamental knowledge to strengthen foundations for more applied studies, technology alternatives not only to support actions but to support capacities for decision-making, and improved tools for analysis. Needs that have been identified include:

Action-oriented knowledge

- Relatively fine-grained climate change projections, especially for the next 20-30 years – e.g., wind regimes for windpower, regional droughts and heat waves
- Improved information about regional implications of water scarcity, along with alternative adaptive responses
- Improved understanding of sensitivities of renewable energy supply systems to changes in climate parameters
- Linkages between energy system adaptation and mitigation
- Contingency planning for vulnerable areas

Fundamental knowledge

- Implications of extreme events for energy system resilience
- Better understanding of coupled systems that include ecological, human-behavior, and technological feedbacks
- Treatments of variance, extremes, and uncertainties: e.g., probabilistic methods, uncertainty quantification
- Non-linearities and tipping points/thresholds as well as performance degradation leading up to abrupt changes

Technology research and development to support energy assessments and actions

- Improved IT systems, including monitoring and control systems to increase information and support flexible responses to disruptive events

- Strategies and technologies to increase resilience and flexibility in electricity supply systems: e.g., greater tolerance for heating, more flexible and smart grids, more distributed generation
- Technologies and policies for electricity peak-shaving
- Materials to cope with new operating conditions, such as heat and ocean acidification

Analytical tools

- Risk management science: risk-based scaling/framing/scoping capabilities, especially given uncertainties that surround large investments for long-term structures
- Improving concepts and tools for modeling integration

B.   *Gaps In Knowledge*

Among these needs for knowledge, several seem to stand out as especially glaring gaps in what we know now as a basis for risk management:

- Relatively fine-grained climate change projections, especially for the next 20-30 years – e.g., wind regimes for windpower, regional droughts and heat waves
- Improved IT systems, including monitoring and control systems to increase information and support flexible responses to disruptive events
- Resilience to extreme events, including interdependencies that can produce major cascading consequences (see NCA technical input report on Climate Change and Infrastructure, Urban Systems, and Vulnerabilities)
- Better understanding of coupled systems that include ecological, human-behavior, and technological feedbacks
- Treatments of variance, extremes, and uncertainties: e.g., probabilistic methods, uncertainty quantification
- Non-linearities and tipping points/thresholds as well as performance degradation leading up to abrupt changes
- Risk management science: risk-based scaling/framing/scoping capabilities, especially given uncertainties that surround large investments for long-term structures

C.   *An example of a need for improved capacities*

While the ability to forecast future climate impacts with spatial and temporal accuracy is still limited, there have been significant improvements in recent years in understanding specific renewable energy technology vulnerabilities to climate change and the technical parameters for operating current technologies efficiently under different environmental scenarios. For example, competition for water resources in key sectors such as agriculture and energy, including biomass feedstock production, other renewables, and thermal generation is likely to become more prevalent under most emissions scenarios (UCS, 2011). As a result, more effort is going toward an improved, and more detailed, understanding of water intensity of both renewable and thermal power technologies (Macknick et al, 2011, Lux Research; 2009). These data and tools can help to inform system level analyses to evaluate the portfolio of RE and thermal capacity in a region and understand how this system can be managed in a climate impacted environment including changes in hydrological cycles and fluctuating RE resource availability. While most of these analyses still focus primarily on existing technologies and storage options, it is anticipated that the future direction of this work will focus on evaluating requirements for various technologies to efficiently operate under more variable environmental parameters including ambient temperature, water availability, water temperature, and humidity. For example, the ability to efficiently operate large scale PV and CSP in desert environments will be impacted not just by the solar resource potential but also negatively affected by the increase and severity of dust and sand storms, higher humidity impacting solar radiation, and humidity effects on module performance and maintenance requirements. Industry is starting to evaluate these types of operational impacts, but there is still little technology and site data to inform how to make these technologies less vulnerable to climate change.

While downscaled climate projections still have serious challenges in accurately representing future wind speeds, frequency distribution, and direction more detailed work is being pursued on these topics (World Bank, 2011, p. 32). An improved level of research and analysis of wind technology vulnerabilities to extreme weather events, including high wind, hail, and icing, has also been carried out in recent years (Pryor and Barthelmie, 2010, 2011a). This research, linked with site-specific empirical data, will be useful in informing local planning.

In the wind sector, there has also been a marked increase in the number of publications on variables affecting the vertical wind profile as well as site specific assessments of climate impacts on wind including wind potential shifts due to moisture and temperature for existing fields and potential for permafrost areas (World Bank, 2011, p. 28; Murphy, 2008). As industry moves toward increased hub height and developers have greater flexibility in siting geographically and at which height, this type of data will be critical. At same time, having clearer data on not just geographic shifts of resource but also localized changes in the wind profile at various hub heights will be important to both project performance for existing capacity and attracting finance for future investments in this sector. While it is true that the shorter life span of a wind installation may make an accurate assessment of these impacts less critical (World Bank, 2011), these farms often have significant sunk costs for permitting, siting, and transmission and potentially storage so there is an economic incentive to fully understand long term wind potential of a given site even if the initial technology may be changed out or retrofitted over time.

A growing number of case studies from U.S., the Andes, and Africa that evaluate power generation potentials related to hydrologic variations are also available

(World Bank, 2011). At the same time, modeling of climate impacts on hydropower production with multiple variables including precipitation spatially and temporally, rate of glacial melt, change in snow pack, and temperature change on reservoir relative to reservoir size, design, and management is extremely complex. With hydropower holding the largest share of the total installed renewables globally, this is a significant challenge and poses a great deal of uncertainty for energy systems – especially in those countries and regions heavily dependent on hydropower for a large share of their total power production.

D.   *Assessment Findings*

Regarding knowledge and research gaps, we find that:

- **Improving knowledge about vulnerabilities and possible risk management strategies is essential for effective climate change risk management in the energy sector**

High consensus, moderate evidence

See Section V A

- **Particularly important is improving knowledge about and improving capacities related to potentials for renewable energy development, resilience to extreme events, and potential tipping points for particular aspects of energy supply and use**

High consensus, moderate evidence

See Section V B

VI.   Toward a Continuing Assessment: Developing the Capacities for National Monitoring, Evaluation, and Informing Decisions about Energy Supply and Use Issues

A.   *Toward a Partnership Approach*

Energy supply and use is a sector distinctively characterized by collaboration in knowledge development and use, despite widespread impressions that different agendas interfere with expert communication. Unlike, say, the United Kingdom, where government officials rarely interact with private sector leaders as peers, in the United States experts from the Department of Energy, EPA, NOAA, and the Department of the Interior can come together in conferences and symposia with experts from the oil and gas and electric utility industries and with experts from non-governmental institutions and academia and exchange knowledge and views remarkably freely.

For example, the annual Energy Modeling Forum in Snowmass, CO, includes representatives of all of these groups, many of whom develop strong personal contact networks that are used actively. EPRI and industry associations hold meetings that bring together all kinds of experts, as do university research centers. Committees and panels of the National Academies of Science/National Research Council included representatives of all of the categories of expertise who work together on consensus statements on important issues. Individuals move back and forth across boundaries between government and non-government.

For energy supply and use, therefore, any self-sustaining long-term structure for continuing climate change consequence assessments will necessarily involve partnerships, not just between knowledge suppliers and users but among all parties as both suppliers and users. What is important is to involve the entire multi-institutional community in clarifying what each kind of institution does best, what kinds of benefits each kind would get from a long-term structure, and how to collaborate in ways that respect aspects of the knowledge base that are proprietary without letting that protection become a barrier to widely useful generic knowledge.

In developing this kind of partnership, which already exists in some respects in an ad hoc but active manner, there are both science issues (i.e., what knowledge the community needs) and institutional issues (i.e., how best to develop and share that knowledge). But a key is likely to be embedding science improvements in value chains: understanding and vitalizing relationships between research, practice, and value for participants.

Among the many aspects of progress are:

- linking research and practice: developing a systematic framework for framing choices, addressing issues regarding what constitute good decisions in the face of uncertainties – perhaps using energy supply and demand as a focus area in this regard within the longer-term NCA structure

- public-private sector partnerships: integrating relevant knowledge from basic research to commercial operations, again with energy as a focus area for NCA, related to real technology R&D and use needs

- providing value at multiple scales: international, national, regional, local, even households, again an opportunity for energy to be a focus area for NCA to explore connections between scales and ways to communicate iteratively with all scales

Aside from relationships with the private sector, a sector that merits particular attention in a self-sustaining continuing assessment process is universities. Institutions of higher learning may be the best prospects to serve as regional hubs for continuing assessment processes as a part of their institutional "brand" because of its value for learning, education, and outreach. Clearly, the federal government recognizes a need for a partnership with universities for climate change assessments and "climate services." For example, NOAA puts its RISAs in regional universities, as does DOI for its regional science centers. Some people suggest land-grant institutions as the best prospects. Others suggest regionally-oriented institutions with both teaching and service-oriented outreach as the best prospects. Still others believe that some of the leading public universities would step forward, especially those with research relationships with industry. In any case, universities have the capacity to make a commitment to a long-term role, assuring appropriate staffing and institutional support, with benefits to their core roles as contributors to knowledge and their linkages with other partners in the nation and their region.

B.   *Challenges in Developing Self-Sustaining Science Based Assessments*

For such a multi-institutional partnership, challenges in improving the science for continuing assessments include:

- strengthening linkages between climate science and energy impact science and practice, especially regarding scenarios

- enhancing scientific capacities for integrated analysis and assessment, including relationships between energy-climate change risks/impacts and other energy policy/practice issues, with attention to model interoperability

- increasing the capacity to acquire emerging knowledge from experience as well as formal published research, including experience from efforts to make infrastructures and urban systems more climate-resilient

- treatments of variance, extremes, and uncertainties, e.g., probabilistic methods, uncertainty quantification, related to risk management

- Improvements in data, especially climate data needed to inform critical energy supply risk management

Challenges in crafting an effective, self-sustaining institutional partnership include:

- Clarifying institutional roles and benefits in filling gaps in the national knowledge base as a national responsibility, not just a federal government responsibility

- Clarifying conditions under which private sector partners can share their knowledge with others

- Deploying for collaborative, iterative monitoring, evaluation, and learning

- Exploring the willingness of an array of universities to take the lead as regional hubs for the partnership

- Establishing a funding mechanism to facilitate continuing institutional relationships and commitments

Aside from relationships with the private sector, a sector that merits particular attention in a self-sustaining continuing assessment process is universities, where the federal government recognizes a need for a partnership with universities for climate change assessments and "climate services" (see above).

C. *Assessment Findings*

Regarding the challenge of developing a self-sustained assessment process for the longer term, we find that:

- **A self-sustaining long-term assessment process needs a commitment to improve the science base, working toward a vision of where things should be in the longer term**

High consensus, moderate evidence	**See Section V B**

- **Capacities for long-term assessments of vulnerabilities, risks, and impacts of climate change on energy supply and use will benefit from effective partnerships among a wide range of experts and stakeholders**

High consensus, moderate evidence	**See Section V A**

- **Self-sustaining assessment structures will provide value to all partners**

High consensus, strong evidence

See Section V A, B

REFERENCES

Ackerman, F. and E. A. Stanton, 2011. *The Last Drop: Climate Change and the Southwest Water Crisis,* Stockholm Environment Institute. Last Accessed on 11-12-2011 at http://www.sei-us.org/publications/id/371

Anada, R., 2011. Midwest Floods: Both Nebraska Nuclear Power Stations Threatened, *Global Research. Accessed at http://globalresearch.ca/index.php?context=va&aid=25307*

Aroonruengsawat A. and M. Auffhammer, 2009. *Impacts of Climate Change on Residential Electricity Consumption: Evidence from Billing Data,* CEC-500-2009-018-F, California Energy Commission, Sacramento, CA. Last Accessed on 11-21-2011 at http://www.energy.ca.gov/2009publications/CEC-500-2009-018/CEC-500-2009-018-F.PDF

Ayers, J. and S. Huq, 2008. The Value of Linking Mitigation and Adaptation: A Case Study, *Environmental Management,* published online Oct 28, 2008, doi:10.1007/s00267-008-9223-2.

Bard, E. and M. Frank, 2006. Climate change and solar variability: What's new under the sun? *Earth and Planetary Science Letters Frontiers.*

Belzer, D. B., M. J. Scott, and R. D. Sands, 1996. Climate Change Impacts on U.S. Commercial Building Energy Consumption: An Analysis Using Sample Survey Data, *Energy Sources* 18(2): 177–201.

Bierbaum, R. 2007. *Coping with Climate Change,* Proceedings of a National Summit, School of Natural Resources and Environment, University of Michigan, Ann Arbor. Last Accessed on 12-02-2011 at http://www.tandfonline.com/doi/abs/10.3200/ENVT.50.4.59-65

Bloom, A., V. Kotroni, and K. Lagouvardos, 2008. Climate change impact of wind energy availability in the Eastern Mediterranean using the regional climate model PRECIS, *Natural Hazards and Earth System Sciences* 8: 1249–1257.

Brooks, F. J., 2000. *GE Gas Turbine Performance Characteristics,* GER-3567H, GE Power Systems, Schenectady, NY, October 2000., Last Accessed on 11-08-2011 at http://www.docstoc.com/docs/79439079/GER-3567H---GE-Gas-Turbine-Performance-Characteristics

Brown, M. A. and B. K. Sovacool, 2011. *Climate Change and Global Energy Security: Technology and Policy Options, MIT Press, 2011.*

Brown, M. A. and M. Dworkin, 2011. "The Environmental Dimension of Energy Security," *Routledge Handbook of Energy Security,* Routledge Press, pp. 176-190.

66

Burkett, V., 2011. Global climate change implications for coastal and offshore oil and gas development, *Energy Policy* 39: 7719–7725.

Burkhardt, G., 2011. Grid on a Hot Streak: Texas sets energy demand record 3 days in a row, *The Monitor,* August 2011.

Burt C., D. Howes, and G. Wilson, 2003. *California Agricultural Water and Electric Energy Requirements, Final Report,* ITRC Report R 03-006, prepared for the California Energy Commission Public Interest Energy Research Program by the Irrigation Training and Research Center, California Polytechnic State University (Cal Poly) San Luis Obispo, CA; Last accessed 11-8-2011 at http://www.itrc.org/reports/energyreq/energyreq.pdf

California Energy Commission (CEC). 1999. *High Temperatures & Electricity Demand: An Assessment of Supply Adequacy in California, Trends & Outlook.Last accessed 11-8-2011 at:* http://www.energy.ca.gov/reports/1999-07-23_HEAT_RPT.PDF

California Energy Commission, 2011. Preliminary California Energy Demand Forecast, 2012-2022. Draft Staff Report, CEC-2000-2011-011-SD, August 2011*. Last accessed at 11-13-2011 at: http://www.energy.ca.gov/2011publications/.../CEC-200-2011-011-SD.pdf*

CBO, 2009.   *The Economic Effects of Legislation to Reduce Greenhouse-Gas Emissions.* Washington, DC: Congressional Budget Office, September 2009.   Last accessed on 10-30-2011 at: *http://www.cbo.gov/sites/default/files/cbofiles/.../09-17-greenhouse-gas.pdf*

Chum, H., et al., 2011. "Bioenergy", in SRREN.

Clayton, M, 2011. Irene leaves 5.5 million without power. Can power companies do better? *The Christian Science Monitor. Last accessed on 11-02-2011 at http://* http://www.csmonitor.com/USA/2011/0829/Irene-leaves-5.5-million-without-power.-Can-power-companies-do-better

Clifton, A. and J. K. Lundquist, 2011. Revealing the Impact of Climate Variability on the Wind Resource Using Data Mining Techniques (Poster), NREL Report No. PO-5000-53526, National Renewable Energy Laboratory. Accessed 01-03-2012 at *http://www.nrel.gov/docs/fy12osti/53526.pdf*

Contreras, S., et al., 2009. Regional Evidence Regarding U.S. Residential Electricity Consumption, Empirical Economics Letters 8: 827–832.

Cooley, H., J. Christian-Smith, and P. H. Gleick, 2008. *More with Less, Agricultural Water Conservation and Energy in California, A Special Focus on the Delta,* Pacific Institute, Oakland, CA. . Last Accessed 11-8-2011 at http:// www.pacinst.org/reports/more_with_less_delta

Cooley, H., et al., 2007. *Hidden Oasis Water Conservation and Efficiency in Las Vegas.* Pacific Institute, Oakland, CA, and Western Resources Advocates, Boulder, CO. Last accessed 10-13-2011 at http://books.google.com/books/about/Hidden_Oasis.html?id=CzZVIgAACAAJ

Cooley, H., J. Fulton, and P. H. Gleick, 2011. *Water for Energy: Future Water Needs for Electricity in the Intermountain West,* Pacific Institute, Oakland CA. Last accessed 11-02-2011 at http://www.pacinst.org/reports/water_for_energy/

Crawley, D. B., 2008. Estimating the impacts of climate change and urbanization on building performance, *Journal of Building Performance Simulation* 1(2): 91–115.

Crowley C., and F. Joutz, 2005. *Weather Effects on Electricity Loads: Modeling and Forecasting,* George Washington University: prepared for EPA. Last accessed on 10-21-2011 at http://*www.ce.jhu.edu.*

Cutter, S., et al., 2008. Community and Regional Resilience*: Perspectives from Hazards, Disasters, and Emergency Management*, CARRI Research Report 1. Oak Ridge, TN: Community and Regional Resilience Institute.

Dell, J. J., and P. Pasateris, 2010. Adaptation in the Oil and Gas Industry to Projected Impacts of Climate Change, Society of Petroleum Engineers , SPE-126307, Last accessed 11-12-2011 at http://www.onepetro.org/mslib/servlet/onepetropreview?id=SPE-126307-MS

De Lucena A., et al., 2009. The vulnerability of renewable energy to climate change in Brazil, *Energy Policy* 37: 879–889.

DOE (U.S. Department of Energy), 2011. *2010 Building Energy Data Book,* last updated March 2011, U.S. Department of Energy, Office of Energy Efficiency and Renewable Energy, Last accessed on 12-5-2011 at http://buildingsdatabook.eren.doe.gov/docs/htm/1.1.4.htm and at http://buildingsdatabook.eren.doe.gov/ChapterIntro8.aspx

EIA (U.S. Department of Energy, Energy Information Administration), 2005. Impacts of Temperature Variation on Energy Demand in Buildings, *Issues in Focus, AEO2005.*

EIA (U.S. Department of Energy, Energy Information Administration), 2006. Manufacturing Energy Consumption Survey. Last Accessed on 11-14-2011. http://www.eia.gov/oiaf/aeo/otheranalysis/aeo_2005analysispapers/vedb.html

EIA (U.S. Department of Energy, Energy Information Administration), 2008. Trends in Heating and Cooling Degree Days: Implications for Energy Demand. *Issues in Focus, AEO2008,* Last accessed on 12-5-2011 at http://www.eia.gov/emeu/mecs/mecs2006/2006tables.html

EIA (U.S. Department of Energy, Energy Information Administration), 2009. *2006 Energy Consumption by Manufacturers – Data Tables,* released June 2009, Table 1.2, Consumption of Energy for All Purposes by Manufacturing Industry and Region (trillion Btu),

EIA (U.S. Department of Energy, Energy Information Administration), 2010. World Energy Outlook, Energy Information Administration, Washington, DC. Ast accessed on 12-01-2011 at http://www.iea.org/weo/

EPRI (Electric Power Research Institute). 2002. *Water & Sustainability (Volume 4): U.S. Electricity Consumption for Water Supply & Treatment—The Next Half Century,* Topical Report 1006787, Electric Power Research Institute, Palo Alto, CA, Last accessed on 12-2-2011 at *www.nist.gov/tip/wp/pswp/upload/243_energy_infrastructure2.pdf*

EPRI (Electric Power Research Institute), 2011. *Water Use for Electricity Generation and Other Sectors: Recent Changes (1985–2005) and Future Projections (2005–2030)*, EPRI Technical Report, Palo Alto, CA. Last accessed on 08-21-2011 at http://my.epri.com/portal/server.pt?space=CommunityPage&cached=true&parentname=ObjMgr&parentid=2&control=SetCommunity&CommunityID=404&RaiseDocID=000000000001023676&RaiseDocType=Abstract_id

EPRI and NERC. 2008. Joint Technical Summit on Reliability Impacts of Extreme Wether and Climate Change, Proceedings . EPRI Report No. 1016095, December 2008. Last accessed on 10-30-2011 at *www.nerc.com/docs/pc/riccitf/EPRI_NERC_PSERC0_1016095.pdf*

ERCOT (Electric Reliability Council of Texas), 2011. Long-*Term Hourly Peak Demand and Energy Forecast,* June 30, 2011. Last Accessed on 11-17-2011. http://www.ercot.com/content/news/presentations/2011/2011_Long-Term_Hourly_Peak_Demand_and_Energy_Forecast.pdf

Fears, D., A., 2011. New Way of Thinking as Sea Levels Rise, *Washington Post*, June 26, 2011, last accessed on 03-12-2012 at http://www.**washingtonpost**.com/...**new-way-of-thinking-as-sea-levels-rise/**...

Faeth, P., 2012. "U.S. Energy Security and Water: The Challenges We Face," Environment, 54/1: 5-19.

FERC and NAERC, 2011. Report on Outages and Curtailments During the Southwest Cold Weather Event of February 1-5, 2011: Causes and Recommendations, Prepared by the Staffs of the Federal Energy Regulatory Commission and the North American Electric Reliability Corporation, August 2011. Last accessed on 2-24-2012 at http://*www.ferc.gov/legal/staff-reports/08-16-11-report.pdf*

Fischer. G., et al., 2007. Climate change impacts on irrigation water requirements: Effects of mitigation, 1990–2080, *Technological Forecasting & Social Change* 74: 1083–1107.

Franco, G. and A. H. Sanstad, 2008. Climate Change and Electricity Demand in California, *Climatic Change* 87(Suppl 1): S139–S151.

GAO (U.S. Government Accountability Office), 2011. *Energy-Water Nexus: Amount of Energy Needed to Supply, Use, and Treat Water Is Location-Specific and Can Be Reduced by Certain Technologies and Approaches,* Report GAO-11-225, US Government Accountability Office, Washington, DC, . Last accessed 12-6-2011 at http://www.gao.gov/new.items/d11225.pdf.

GE Energy, 2010. Western wind and solar integration study, prepared for the National Renewable Energy Laboratory, available through http://www.osti.gov/bridge

Gleick, P. H, et al., 2003. *Waste Not, Want Not: The Potential for Urban Water Conservation in California,* Pacific Institute, Oakland, CA. 176 pp,

Gleick, P. H. and H. S. Cooley, 2009. Energy Implications of Bottled Water, *Environmental Research Letters* 4: 014009, doi: 10.1088/1748-9326/4/1/014009.

Greene, D. L., et al., 2010. The Importance of Advancing Technology to America's Energy Goals," *Energy Policy*, 38 (2010): 3886-3890.

Gueymard, C. A., and S. M. Wilcox, 2011. Assessment of Spatial and Temporal Variability in the U.S. Solar Resource from Radiometric Measurements and Predictions from Models Using Ground-Based or Satellite Data, *Solar Energy* 85(5): 1068–1084; NREL Report No. JA-5500-49849; doi: 10.1016/j.solener.2011.02.030, last accessed on 12/01-2011 at *http://www.sciencedirect.com/science/article/pii/S0038092X11000855*

Haber, H., et al., 2011. Global bioenergy potentials from agricultural land in 2050: Sensitivity to climate change, diets and yields, *Biomass and Bioenergy* 35(12).

Hamlet, A. F., et al., 2009. Effects of projected climate change on energy supply and demand in the Pacific Northwest and Washington State, in Chapter 4, Energy, of *The Washington Climate Change Impacts Assessment: Evaluating Washington's Future in a Changing Climate,* JISAO Climate Impacts Group, University of Washington, Seattle. Last accessed on 12-18-2011 at http://cses.washington.edu/cig/res/hwr/ccenergy.shtml

Hamlet, A. F., 2010. Effects of projected climate change on energy supply and demand in the Pacific Northwest and Washington State. *Climatic Change* 102:103–128. DOI 10.1007/s10584-010-9857-y

Harrison, G. P., and H. W. Whittington, 2002. Susceptability of the Batoka Gorge hydroelectric scheme to climate change, *J. Hydrol.* 264(1–4): 230–241.

Hayhoe, K., et al., 2010. An integrated framework for quantifying and valuing climate change impacts on urban energy and infrastructure: A Chicago case study. *Journal of Great Lakes Research* 36 (S2): 94-105 http://www.sciencedirect.com/science/article/pii/S0380133010000547

Heath, G. A., J. J. Burkhardt, and C. S. Turchi, 2011. *Life Cycle Assessment of a Parabolic Trough Concentrating Solar Power Plant and Impacts of Key Design Alternatives: Preprint,* NREL Report No. CP-6A20-52186, National Renewable Energy Laboratory, 11 pp. Last accessed 0n 11-21-2011 at http://nrelpubs.nrel.gov/Webtop/ws/nich/www/public/Record;jsessionid= F0739E4738EF12073EE1DD0DC2E6C9AC?rpp=25&upp=0&m=1&w=NATIV E%28%27AUTHOR+ph+words+%27%27%22heath%2Bg%22%27%27%27 %29&order=native%28%27pubyear%2FDescend%27%29

Hekkenberg, M., H. C. Moll, and A. J. M. Schoot Uiterkamp, 2009. Dynamic temperature dependence patterns in future energy demand models in the context of climate change, *Energy* 34: 1797–1806.

IPCC, 2007. *Climate Change 2007: Impacts Adaptation, and Vulnerability.* Working Group II Contribution to the Fourth Assessment Report of the Intergovernmental Panel on Climate Change. New York: Cambridge University Press.

Isaac, M., and D. P. van Vuuren, 2009. Modeling global residential sector energy demand for heating and air conditioning in the context of climate change. *Energy Policy* 37: 507–521.

Jo, J. H., et al., 2010. Sustainable urban energy: Development of a mesoscale assessment model for solar reflective roof technologies, *Energy Policy* 38: 7951–7959.

Kates, R., W. Travis, and T. Wilbanks, 2012. "Transformational Adaptation when Incremental Adaptations to Climate Change Are Insufficient," submitted to *Proceedings of the National Academies of Science (PNAS),* 2012

Kawajiri, K., T. Oozeki, and Y. Genchi, 2011. Effect of Temperature on PV Potential in the World, *Environmental Science & Technology,* 45: 9030–9035.

Kenney, D. S., and R.Wilkinson, 2011. *The Water-Energy Nexus in the American West,* Edward Elgar, Northampton, MA, 296 pp.

Kenny, J. F., et al. 2009. *Estimated Use of Water in the United States in 2005,* Circular 1344, U.S. Geological Service, Reston, VA. Last accessed 02-03-2012 at http://pubs.usgs.gov/*circ/1344/pdf/c1344.pdf*

Kiameh P., 2003. *Power Generation Handbook,* McGraw-Hill, New York, p. 4.8.

Klein, G. 2005. *California's Water–Energy Relationship, Final Staff Report,* CEC-700-2005-011-SF, prepared in support of the 2005 Integrated Energy Policy Report Proceeding (04-IEPR-01E), California Energy Commission, Sacramento, CA. Last accessed 10-21-2011 at http://*www.fred.ifas.ufl.edu/.../Paul-Lander-references-webinar-4-12-2011..*

Kopytko, N., and J. Perkins, 2011. Climate Change, Nuclear Power, and the Adaptation-Mitigation Dilemma, *Energy Policy* 39: 318–333.

Kurtz, S., et al., 2009. Evaluation of High-Temperature Exposure of Rack-Mounted Photovoltaic Modules, pp. 002399–002404 in *Proceedings 34th IEEE Photovoltaic Specialists Conference (PVSC '09),* June 7-12, Philadelphia, NREL Report No. CP-520-45644, Institute of Electrical and Electronics Engineers, Piscataway, NJ. Last accessed 11-23-2011 at http://ieeexplore.ieee.org/xpls/abs_all.jsp?arnumber=5411307&tag=1

Levinson, R., and H. Akbari, 2010. Potential benefits of cool roofs on commercial buildings: conserving energy, saving money, and reducing emission of greenhouse gases and air pollutants, *Energy Efficiency* 3: 53–109, DOI: 10.1007/s12053-008-9038-2.

Lu, N., et al., 2010. Climate Change Impacts on Residential and Commercial Loads in the Western U.S. Grid, *IEEE Transactions on Power Systems* 25(1): 480–488.

Lux Research, 2009. Global energy: Unshackling carbon from water. LRWI-R-09-03. Last accessed 02-23-2012 at https://portal.luxresearchinc.com/research/document_excerpt/5035.

MacDonald, G. M., 2010. From the Cover: Climate Change and Water in Southwestern North America Special Feature: Water, Climate Change, and Sustainability in the Southwest, *Proceedings of the National Academy of Sciences* 107(50): 21256–21262.

Macknick, J., et al., 2011. *Review of Operational Water Consumption and Withdrawal Factors for Electricity Generating Technologies,* NREL Report TP-6A20-50900, 29 pp. Last accessed on 01-27-2012 at http://*www.nrel.gov/docs/fy11osti/50900.pdf*

Markoff, , M.S. and A.C. Cullen, 2008. Impact of Climate Change on Pacific Northwest Hydropower. Climatic Change 87: 451-469

McNei,l M.A. and V. E. Letschert, 2007. Future air conditioning energy consumption in developing countries and what can be done about it: the potential of efficiency in the residential sector, Paper 6306, Lawrence Berkeley National Laboratory, Last accessed on 11-17-2011 at http://escholarship.org/uc/item/64f9r6wr

Messner, S. , et al., 2009. *Climate Change-Related Impacts In The San Diego Region By 2050.* CEC-500-2009-027-F. A Paper from the California Climate Change Research Center, last accessed on 11-20-2011 at http*://www.energy.ca.gov/**2009***publications/...***2009**.../CEC-500-**2009**-027.*

Mideska, T. K., and S. Kallbekken, 2010. The impact of climate change on the electricity market : A review, *Energy Policy* 38: 3579–3585.

Miller, N. L., J. Jin J., K. Hayhoe, and M. Auffhammer, 2007. *Climate Change, Extreme Heat, and Electricity Demand in California,* PIER Project Report CEC-500-2007-023, prepared for the California Energy Commission Public Interest Energy Research Program, Sacramento, CA., Last accessed on 10-15-2011 at http*://www.fypower.org/pdf/CEC_CC-ElectricityDemand.PDF*

Miller, N. L., J. Jin, K. Hayhoem and M. Auffhammer, 2008. Climate Change, Extreme Heat, and Electricity Demand in California, *Journal of Applied Meteorology and Climatology* 47: 1834–1844.

Nakićenović, N., et al., 2000. *IPCC Special Report on Emissions Scenarios,* Intergovernmental Panel on Climate Change, Cambridge University Press, Cambridge.

New York State Energy Planning Board, 2011. *Scope for the 2013 New York State Energy Plan*, Last Accessed on 11-21-2011 at http://www.nysenergyplan.com/meeting/Scope%20for%20the%202013%20Energy%20Plan.pdf.

Novotny, V., 2010. Urban Water and Energy Use from Current US Use to Cities of the Future, pp. 118–140 in *Proceedings, Cities of the Future/Urban River Restoration 2010,* Water Environment Federation, Alexandria, VA, Last accessed on 12-6-2011 at http://aquanovallc.com/wp-content/uploads/2010/09/URBAN-WATER-AND-ENERGY-USE.pdf

NPCC (Northwest Power and Conservation Council), 2010. *Sixth Northwest Power Plan, Appendix L. Climate Change and Power Planning,* Report 2010-09, Northwest Power and Conservation Council, Portland, OR. Last Accessed on 11-15-2011 at http://www.nwcouncil.org/energy/powerplan/6/final/SixthPowerPlan_Appendix_L.pdf.

NRC (National Research Council), 2010. *Adapting to the Impacts of Climate Change,* Washington, DC.

NRC (National Research Council). 2011. *America's Climate Choices,* Washington, DC.

NRC (National Research Council), 2009. America's Energy Future: Technology and Transformation, . ISBN-10: 0-309-11602-3, National Academies Press, Washington, DC, 650 pp.

Pacific Institute, 2011: Water for Energy: Future Water Needs for Electricity in the Intermountain West, Pacific Institute, Oakland, CA Last accessed on 01-14-2012 at http://www.pacinst.org/reports/water_for_energy/

Perez, P., et al., 2009. *Potential Impacts of Climate Change on California's Energy Infrastructure and Identification of Adaptation Measures,* California Energy Commission Staff Paper CEC-150-2009-001. Last Accessed 11-21-2011 at http://www.energy.ca.gov/2009publications/CEC-150-2009-001/CEC-150-2009-001.PDF

Persson, T., et al., 2009. Maize, ethanol feedstock production and net energy value as affected by climate variability and crop management practices, *Agricultural Systems* 100: 11–21.

PNNL/DOE (Multiple Authors), 2008. *Climate Change Impacts on Residential and Commercial Loads in the Western U.S. Grid.*, PNNL-17826. Last accessed on 11-18-2011 at http://www.pnl.gov/publications/abstracts.asp?report=251787

Poudel, B. C., et al. , 2011. Effects of climate change on biomass production and substitution in north-central Sweden, *Biomass and Bioenergy* 35(10).

Proceedings of the Workshop on Electric Utilities and Water: Emerging Issues and R&D Needs, July 23–24, 2002, Pittsburgh, PA, U.S. Department of Energy, Office of Fossil Energy, National Energy Technology Laboratory. Last accessed on 11-28-2011 at https://docs.google.com/viewer?a=v&q=cache:sYRTtjnGwPEJ:www.netl.doe.gov/publications/proceedings/02/EUW/WEF%2520Paper%2520Final%2520header.pdf+Proceedings+of+the+Workshop+on+Electric+Utilities+and+Water:+Emerging+Issues+and+R%26D+Needs&hl=en&gl=us&pid=bl&srcid=ADGEESg2EcHvVsUQ4NKzKqUC4fblRZzRHSvpBXdNXCNhMaaIuauxOyVQNoZBsnOed3H0Imq7-sdZ45Aoi2QJlMFbfkv1IUxMX6xr6H7fhSms5HEQZewpzszUWshzt3fY_thldYN-Idn7&sig=AHIEtbQ-wlUNXJVOahdNejw2AEYh2K09Uw

Pryor, S. C. and R. J. Barthelmie, 2010. Climate change impacts on wind energy: A review, *Renewable and Sustainable Energy Reviews* 14.

Pryor, S. C. and J. Ledolter, 2010. Addendum to "Wind speed trends over the contiguous United States," *J. Geophys. Res.* 115: D10103, doi: 10.1029/2009JD013281.

Pryor S. C. and R. J. Barthelmie, 2011. Assessing climate change impacts on the near-term stability of the wind energy resource over the United States, *Proceedings of the National Academy of Sciences*, doi: 10.1073/pnas.1019388108.

Rong, F., 2006. *Impact of Urban Sprawl on U.S. Residential Energy Use.* PhD Dissertation, School of Public Policy, University of Maryland, College Park, MD, last accessed 11-8-2011 at umi-umd-3694.pdf.

Rosenzweig, C. et al., 2006. Assessing potential public health impacts of changing climate and land uses: The New York Climate and Health Project. In *Regional Climate Change and Variability*. M. Ruth, K. Donaghy, and P. Kirshen, Eds., New Horizons in Regional Science. Edward Elgar.

Rosenzweig, C., W. D. Solecki, and R. B. Slosberg, 2006. *Mitigating New York City's Heat Island with Urban Forestry, Living Roofs, and Light Surfaces.* NYSERDA Report 06-06, 133 pp.

Rosenzweig, et al., 2009: Mitigating New York City's heat island: Integrating stakeholder perspectives and scientific evaluation. *Bull. Amer. Meteorol. Soc.*, **90**, 1297-1312, doi:10.1175/2009BAMS2308.1. Last accessed 03-16-2012 at http://www.epa.gov/heatisld/resources/reports.htm#mitigation

Ruth, M. (ed), 2006. *Smart Growth and Climate Change,* Edward Elgar, Northampton, MA, 409 pp.

Sailor, D., 2001. Relating residential and commercial sector electricity loads to climate – evaluating state level sensitivities and vulnerabilities, *Energy* 26: 645–657.

Sailor, D. J. and A. A. Pavlova, 2003. Air Conditioning Market Saturation and Long-Term Response of Residential Cooling Energy Demand to Climate Change, *Energy* 28(9): 941–951.

Sailor, D. J., M. Smith and M. Hart, 2008. Climate change implications for wind power resources in the Northwest United States, *Renewable Energy* 33(11): 2393–2406.

Sailor, D. and R. Munoz, 1997. Sensitivity of electricity and natural gas consumption to climate, *Energy* 22(10): 1997.

SAP 4.5, 2008. *Effects of Climate Change on Energy Production and Use in the United States,* a report by the U.S. Climate Change Science Program and the Subcommittee on Global Change Research, T. Wilbanks et al. (eds), U.S. Department of Energy.

SAP 4.7, 2008. *Impacts of Climate Variability and Change on Transportation Systems and Infrastructure – Gulf Coast Study.* U.S. Department of Transportation and US Geological Survey

Sathaye J., et al., 2011. *Estimating Risk to California Energy Infrastructure from Projected Climate Change,* California Energy Commission, Publication No. CEC-500-2011-xxx. Last accessed on 03-02-2012 at http://escholarship.org/uc/item/14r3v942.

Scott , C. A., et al., 2011. Policy and institutional dimensions of the water–energy nexus, *Energy Policy* 39(10): 6622–6630, ISSN 0301-4215, 10.1016/j.enpol.201108.013.

Scott, M. J., J. A. Dirks, and K. A., 2008. The Value of Energy Efficiency Programs for US Residential and Commercial Buildings in a Warmer World, *Mitigation and Adaptation Strategies for Global Change* 13: 307–339.

Shorr, N., R. G. Najjar, A. Amato, and S. Graham, 2009. Household heating and cooling energy use in the northeast USA: comparing the effects of climate change with those of purposive behaviors, *Climate Research* 39: 19–30, doi: 10.3354/cr00782.

Souder, E., et al., 2011. Freeze knocked out coal plants and natural gas supplies leading to blackouts, *The Dallas Morning News*, Dallas, TX, February 6, 2011. Last accessed on 11-16-2011 at http://www.dallasnews.com/news/state/headlines/20110206-freeze-knocked-out-coal-plants-and-natural-gas-supplies-leading-to-blackouts.ece

SREX, 2011. *IPCC Special Report on Managing the Risks of Extreme Events and Disasters to Advance Climate Change Adaptation: Summary for Policymakers,* IPCC, Geneva.

SRREN, 2011. *IPCC Special Report on Renewable Energy Sources and Climate Change Mitigation: Summary for Policymakers,* IPCC, Geneva.

Stillwell, A. S., et al.. 2011. The Energy-Water Nexus in Texas, *Ecology and Society* 16(1): 2

Stokes, J. R. and A. Horvath, 2009. Energy and Air Emission Effects of Water Supply, *Environmental Science and Technology* 43: 2680–2387.

Tonn, B. and J. Eisenberg, 2007. The Aging US Population and Residential Energy Demand, *Energy Policy* 35: 743–745.

TVA (Tennessee Valley Authority), 2011. TVA Statement on Climate Change Adaptation. . Last Accessed on 11-15-2011 at http://www.tva.gov/environment/sustainability/climate_change_statement.pdf

UK Climate Change Risk Assessment, 2012. *Climate Change Risk Assessment,* UK
    Department for Environment, Food, and Rural Affairs (DEFRA), London. Last
    accessed on 02-14-2012 at
    http://issuu.com/openbriefing/docs/ccraevidence

Union of Concerned Scientists, 2011. *Freshwater Use by U.S. Power Plants:*
    Electricity's Thirst for a Precious Resource, Energy and Water for a Warming
    World Initiative (EW3). Last accessed on 03-16-2012 at
    http://www.ucsusa.org/clean_energy/technology_and_impacts/impacts/fre
    shwater-use-by-us-power-plants.html

U.S. Fish & Wildlife Service, 2011. Chincoteague National Wildlife Refuge:
    Comprehensive Conservation Planning Update, Chincoteague Island, VA,
    August 2011. Last accessed on 12-15-2011 at
    http://www.fws.gov/northeast/chinco/

U.S. Global Change Research Program, 2009. Global Climate Change Impacts in the
    United States, Cambridge University Press,

Vano, J. A., et al., 2010. Climate change impacts on water management and irrigated
    agriculture in the Yakima River Basin, Washington, USA, *Climatic Change*
    102: 261–286, DOI: 10.1007/s10584-010-9846-1.

Vine, E., 2008. Adaptation of California's Electricity Sector to Climate Change, Public
    Policy Institute of California, supporting report for *Preparing California for a*
    Changing Climate. Last Accessed 11-21-2011 at
    http://www.ppic.org/content/pubs/report/R_1108EVR.pdf

Vine, E., 2011. Adaptation of California's electricity sector to climate change,
    Climatic Change, published online October 6, 2011. Last Accessed 11-21-
    2011 at http://www.ppic.org/content/pubs/report/R_1108EVR.pdf

Washington DOE (Washington Department of Ecology), 2009. Washington Climate
    Change Impacts Assessment 2009, University of Washington Climate Impacts
    Group. Last Accessed on 11-21-2011 at
    http://www.ecy.wa.gov/climatechange/ipa_resources.htm. The study also
    may be found at http://cses.washington.edu/db/pdf/wacciareport681.pdf

Western Governors Association, 2010. Climate Adaptation Priorities for the Western
    States, Last accessed on 11-18-2011 at
    https://docs.google.com/viewer?a=v&q=cache:VY7nz1SwoRYJ:www.westgo
    v.org/component/joomdoc/doc_download/1279-climate-adaptation-
    report+Western+Governors+Association,+Climate+Adaptation+Priorities+fo
    r+the+Western+States,+2010&hl=en&gl=us&pid=bl&srcid=ADGEESjAKhlDw
    5f58xiqIiAPT1xGSanFwcEFfQlM6mFEVARzuMQtz-bgEpjWqiC4g-
    LMjw7PScLvHxRjSmBjVZD1ljGeP7gwgImks8vmX1qwgmhHSB-tqTJBM-
    4_O6G10viulk9dA0Sx&sig=AHIEtbSd7X7WM9LZYKb7Rkra7bSfhYEwWA

Wisconsin DNR (Wisconsin Department of Natural Resources), 2011. *Wisconsin's Changing Climate: Impacts and Adaptation 2011, Wisconsin Initiative on Climate Change Impacts,* Nelson Institute for Environmental Studies, University of Wisconsin-Madison and the Wisconsin Department of Natural Resources, Madison, last accessed on 01-27-2011 at *www.wicci.**wis**c.edu/publications.php*

Wilbanks, T., et al., 2007. Toward an Integrated Analysis of Mitigation and Adaptation: Some Preliminary Findings, in T. Wilbanks, R. Klein, and J. Sathaye (eds), special issue, *Mitigation and Adaptation Strategies for Global Change* 12(5): 713–725.

Wilbanks T. and J. Sathaye, 2007. Integrating Mitigation and Adaptation as Responses to Climate Change: A Synthesis, in T. Wilbanks, R. Klein and J. Sathaye (eds), special issue, *Mitigation and Adaptation Strategies for Global Change*, 12(5): 957–962.

Wilbanks, T., 2010. Research and Development Priorities for Climate Change Mitigation and Adaptation, pp. 77–99 in R. Pachauri (ed), *Dealing with Climate Change: Setting a Global Agenda for Mitigation and Adaptation*, TERI, New Delhi.

Wise, M., et al., 2009. Implications of limiting CO_2 concentrations for land use and energy, *Science*, 324:1183-1186.

World Bank, 2011. *Climate Impacts on Energy Systems: Key Issues for Energy Sector Adaptation,* Energy Sector Management Assistance Program (ESMAP). Washington, DC., last accessed on 02-18-2012 at http://www.esmap.org/esmap/node/1152

Xu, P., et al., 2009. *Effects of Global Climate Changes on Building Energy Consumption and Its Implications on Building Energy Codes and Policy In California*. PIER Final Project Report. CEC-500-2009-006. Prepared for the California Energy Commission Public Interest Energy Research Program, Sacramento, CA. Last accessed on 02-19-2012 at http://www.energy.ca.gov/2009publications/CEC-500-2009-006/CEC-500-2009-006.PDF